SUSTAINABLE DIPLOMACY

SUSTAINABLE DIPLOMACY

ECOLOGY, RELIGION, AND ETHICS IN MUSLIM–CHRISTIAN RELATIONS

David Joseph Wellman

SUSTAINABLE DIPLOMACY
© David J. Wellman, 2004.

First published 2004 by
PALGRAVE MACMILLAN™
175 Fifth Avenue, New York, N.Y. 10010 and
Houndmills, Basingstoke, Hampshire, England RG21 6XS
Companies and representatives throughout the world

PALGRAVE MACMILLAN is the global academic imprint of the Palgrave Macmillan division of St. Martin's Press, LLC and of Palgrave Macmillan Ltd. Macmillan® is a registered trademark in the United States, United Kingdom and other countries. Palgrave is a registered trademark in the European Union and other countries.

ISBN 1–4039–6442–4 hardback

Library of Congress Cataloging-in-Publication Data
Wellman, David J.
    Sustainable diplomacy : ecology, religion, and ethics in Muslim-Christian relations / David Joseph Wellman.
        p. cm.
    Includes bibliographical references (p.) and index.
    ISBN 1–4039–6442–4
    1. Religion and state. 2. Religion and culture. 3. East and West.
    4. Morocco—Relations—Spain. 5. Spain—Relations—Morocco.
    I. Title. II. Series.

BL65.S8W45 2004
327—dc22   2003065612

A catalogue record for this book is available from the British Library.

Design by Newgen Imaging Systems (P) Ltd., Chennai, India.

First edition: April 2004
10 9 8 7 6 5 4 3 2 1

Printed in the United States of America.

For Andy, with love and gratitude

CONTENTS

ACKNOWLEDGMENTS

This book would not exist without an extraordinary community of people who continue to teach me through their patience and generosity. I must begin by thanking my long-time mentor, Larry Rasmussen, who is my standard, and who guided me from the beginnings of this book to its conclusion. I am also deeply grateful for the wisdom of Chung Hyun Kyung, who taught me to celebrate the beauty of inviting everyone to the table. This effort would also not be possible without the insight and patience of Andrea Bartoli, who understands and lives out the art that is conflict resolution. In addition, I would not have been able to engage Islam in any authentic way without the generous guidance and insight of Scott Kugle. Finally, I am deeply indebted to Douglas Johnston, Cynthia Sampson, Edward Luttwak, Barry Rubin, and Stanton Burnett for their collective contributions to the groundbreaking volume entitled *Religion, the Missing Dimension of Statecraft* (New York: Oxford University Press, 1994). Their writings gave pivotal inspiration and guidance to my own work.

Throughout the research for this project, many of the most important insights I gained were acquired during my fieldwork. I am overwhelmingly indebted to the countless Spaniards and Moroccans who took me into their homes and fed me both intellectually and physically in a manner I could not have found in the finest library or restaurant. The gift of their time and their graciousness remains the greatest source of my hope for the work I seek to engage. I want to begin by thanking my Spanish brother, Javier Martos Cando of Seville for his intellectual insight and willingness to help me gather interviewees. I am also deeply grateful for the love and encouragement of the feminist scholar Rosalia Romero Perez, who opened her home to me, and shared with me both her family and her keen desire to understand and explain her Spanish and Moroccan neighbors. There are so many others in Spain without whom my work would have been impossible. They include Lola Villar Gallego, Susana Ricca Ruiz and Louis Ladebauphe Garriga, Lola Caro Sanchez and Sonia Echarri Carrillo, Ana Maria Echarri Carillo and Francisco Castillo, Peter Huijing and Mercedes Carretero Ortego, Casimira Perez Dominguez, Leonardo Savignano, Rafeael Carretero Ortega, Juan Francisco Navarro Escobar, Sebastian de la Obra, and Deborah Avery.

My research in Morocco would never have occurred had it not been for the extraordinary generosity of my Moroccan brother Kamal Hassani Bouzidi Benmajdoub, his father Absalam, his mother Nouzha, his brothers Muhammad and Yusef, and his sister Fatima Zahara. They welcomed me into their home and invited me to live with them as I collected interviews and tried to improve my Arabic. While the Bouzidi's made me a member of their family, many other wonderful Moroccans shared their time, insights, and talents with me. These include Mostafa Ouajjani of Fes, who set me on an extraordinary path. My deep thanks as well goes to Mostafa's brother, Brahim Ouajjani of Tiznit, who introduced me to the South. Many of my interviews would never have occurred without the tireless work of Muhammad Oukili. There are so many others who opened extraordinary doors for me. They include Sadik Rddad, Ibrahim Dhaiba, and Salam Bouzidi.

While the names mentioned above begin to tell the story of how I came to write about Spain and Morocco, mere names do not begin to tell the stories of all that these kind people shared with me. These people made my subject matter come alive, from walking with a romeria procession from Quatravitas and eating olives with a farmer from the village, to riding on the back of a scooter in the Fassi sun to interview an artisan in the brass souk of Fes al-Bali. It is for this reason that I cannot claim to be dispassionate about the people I attempt to describe in the course of this project. Each of them have given me a gift I will never forget: a gift of time, of honor, and of inviting me to their tables. It is my greatest hope that what each of these people have offered me is reflected in some small way in the following pages.

Finally, I could not possibly have accomplished any of my work without the support of a number of Americans, first and foremost being my family. I am so grateful for the tireless patience and guidance of my life partner, Andy Robinson. Nothing would be possible without the extraordinary encouragement and support of my parents, Ann and Norbert Wellman, my sister Kaie Wellman, and my brother-in-law Kevin de Garmo. No less important are the gifts of my in-laws, Jack and Judy Robinson and my other brother-in-law, Chris Robinson. Last, but certainly not least, I very much wish to thank four extraordinary people who had the patience to proofread this text and guide me through the construction of Ecological Footprints: Bill Bliss, John Domini, and Nova and Alyssa Gutierrez.

It should be quite clear by now that I have no right to take sole credit for this book. My work has emerged from a community of people who have been generous enough to share the best of themselves with me. It is my hope that in turn, this work can somehow honor them.

David Wellman
New York, July 2003

INTRODUCTION: AN OVERVIEW

A mong policy makers and scholars of foreign affairs, the predominant form of analyzing relations between nation-states has long been through the lens of a *Realist* worldview. An approach that views all nation-states' political interactions as being motivated exclusively by the desire to acquire, retain, and project power, the Realist view of international relations is based on a broad set of secularist–materialist assumptions regarding human exchanges.[1] While a Realist analysis can provide many remarkable insights as to the motivations of nation-states in the context of economic and military competition and conflict, it is less capable of anticipating, and less inclined to privilege, the impact of individual human motivations, religious and social movements, and identity-based politics on the conduct of international relations.

A religious analysis of international relations widens the origins of nation-state behavior to include not only policy formation based on economic and geostrategic considerations, but also other critical factors.[2] Such factors are found within a nation's modern religious culture, which is influenced by its spiritual history, and the structure of belief of *the people*, and not simply the institutions, which make up a country's leadership as well as its constituency. An exploration and analysis of the religious culture in a nation-state will thus play the central role in determining the parameters of a religious analysis of international relations.

For the purposes of this book, the term "religious culture" will refer to a particular dimension of the social milieu in which all people live, and that is most often distinguished by geographic locale, ethnicity, and nationality. Religious culture teaches people to use language, metaphors, and appeals to moral and ethical norms that are drawn from the dominant religious traditions of their particular geographic locale, but that for many have ostensibly become secularized. As a political phenomenon, religious culture comes into play in the following contexts: (1) in the use of religious symbols or language by a national government or other actors to convey particular meaning or justify supposedly secular actions to its own general populace or other international actors; (2) through religious language and imagery as a vehicle for conveying meaning and value between members of the general population; (3) as an appeal by the state and individuals to moral and ethical norms drawn from what were originally religious sources (particularly, but not limited to, the dominant religious tradition of the respective nation-state); (4) the cultivation by national leadership of the perception that the state acts in concert with, or out of sincere respect toward, the dominant religious institutions and traditions in the nation-state; and (5) the governmental use of religious ties and traditions to fortify the legitimacy of the state leadership and apparatus in the eyes of the people.

This book will assert that by acknowledging the intimate tie between religion and power, a clearer understanding of nation-state conduct, as well as various potential means of improving transnational cooperation, can be attained. To separate the secular from the religious, as most international relations analysis prefers to do, is to block an important angle of interpretation regarding the motivation and behavior of national leaders. It is also probable that in separating religion from political power, the observer runs the risk of not seeing what is taking place on the ground. As the anthropologist Henry Munson, Jr. has noted:

> The relation between religion and power is invariably distorted when we focus exclusively on the overtly political aspects of religion or the overtly religious aspects of power. Without some idea of how a religion is understood by ordinary people in their everyday lives, we cannot begin to assess the political impact of the religious rhetoric of rulers and rebels. Conversely, if we restrict our attention to the overtly religious facets of power, we exaggerate their significance and ignore others—like force, fear, and the rage of people who cannot find work.[3]

Thesis/Questions

In light of the limitations of the Realist approach, this book will argue the critical importance of employing a religious analysis of international relations, as an indispensable addition to existing approaches for interpreting and improving relations between nation-states. To this end, this book will propose a method for constructing a religious analysis of international relations with the specific goal of providing *a new means of identifying the theo-ethical motivations of a nation-state and its population in its relations across international borders*. This goal will be pursued by applying this new method of religious analysis to a particular case study: the modern relationship between Spain and Morocco.

The objective of this book is informed by the central hypothesis that many of the theo-ethical norms that affect and guide the life of a nation's people (including its leaders) can be identified through observing a nation-state's religious traditions through the hermeneutic of the land it occupies. To accomplish this task, the modern analyst of international affairs must examine the "Ecological Location" of the nation's population and its "Ecological Footprint."

A creation of the Christian ethicist Daniel Spencer, Ecological Location is a means of examining the human relationship with the greater Creation, including the human's relationship with the land he or she inhabits and/ or controls.[4] By illumining the human relationship to greater Creation, Ecological Location provides other insights into the norms guiding human conduct. These insights include new approaches to understanding community formation, relationships with those identified as "other," and the words and phrases people use to describe how human relationships with human and nonhuman members of Creation are established, understood, and maintained. Such social and linguistic phenomena not only describe what a people need and do in order to survive, but also what the religious culture in their respective countries has taught them to believe about the origin, purpose, and value of the human and nonhuman world. By identifying and understanding the Ecological Location of one's own nation, as well as in the country or countries with which it is in dialogue, a diplomat comes into possession of an important means of communication and bridge-building. For by determining the common points of agreement within the Ecological Locations of two different nation-states, one may identify common ground for cooperation that translates across differences of nationality, religion, ethnicity, and culture. Simultaneously, Ecological Location can be a highly insightful means of beginning to expose the roots of conflict and the nature of inequalities that exist within and between communities.

An equally important tool to analyze the case study is found in Mathis Wackernagel and William Rees' *Ecological Footprint*. Wackernagel and Rees have created a way to measure the impact of an individual, a community, or even a nation upon Earth's biosphere. By determining the amount of hectares of land a nation uses in generating the amount of natural resources it consumes, Wackernagel and Rees' analysis calculates how far beyond its borders a nation must go in order to sustain its levels of production, consumption, and pollution. While one nation might live within the means of its own frontiers, another may have an Ecological Footprint that stretches around the globe. The use of the Ecological Footprint in gauging the relationship between two neighboring nation-states or bioregions is invaluable, as it demonstrates not only each nation or region's level of dependency upon the other, but also the degree to which their existing ecological relationship provides an avenue for conflict resolution and cooperation.

This book will also introduce two new concepts into international relations discourse: Ecological Realism and Sustainable Diplomacy. Underlying both of these ideas is the notion of approaching diplomacy as an ecological discourse. Ecological Realism is the philosophical counterpoint to traditional Realism. In contrast to the anthropocentric assumptions of Realism, Ecological Realism is eco-centric. While Realism focuses on the centrality of military and economic might, Ecological Realism provides new definitions of power that center on the nation-state's and the bioregion's ability to ecologically sustain themselves. Ecological Realism argues that the Earth economy is an indispensable arbiter of value, and that human monetary systems do not reflect the realities of resource scarcity or the limits of the natural world.

The fate of humans and nonhumans are inherently linked together through our common location within the Earth's biosphere. Thus, power in the context of Ecological Realism focuses on the ability of communities to control their own levels of consumption. Societies are either sustainable or not sustainable. Under Ecological Realism, survival is no longer isolated to one group; it is a collective global goal and an unmitigated good, as all human and nonhuman creatures share the reality of one common fate. For this reason, Ecological Realism requires that the practitioners of diplomacy move beyond an individualistic nation-state to nation-state dialogue and work toward a systemic multilevel approach to foreign relations that focuses on promoting the health of bioregions. Under the rubric of Ecological Realism, either we must all come together to acknowledge the inherent limitations of the biosphere, to act in accordance with the realities of Earth's capacity to support human consumption, or perish together by denying these truths.

Sustainable Diplomacy converts Ecological Realism into policy and practice. To this end, Sustainable Diplomacy maps a terrain that links religion, land, and power in order to analyze the conduct of human communities and their relationship to the biosphere. Sustainable Diplomacy is not only interested in fomenting better relationships between heads of state; rather, it also aims to promote better long-term relations between national populations. This calls for a more intimate and profound understanding of the lives, beliefs, and concerns of people "on the ground."[5] Therefore Sustainable Diplomacy requires willingness to abandon old perceptions and embrace new means of communication. A practitioner of Sustainable Diplomacy must be familiar with the religious beliefs of the populations he or she is engaging, their relationship with the land they live on, and the relationships they carry on with those they call "neighbor." For this reason, Sustainable Diplomacy embraces a systemic approach to international affairs, one in which there is authentic room for NGOs, religious organizations, and various peoples' movements to contribute to policy formation and act as diplomats in their own right. Sustainable Diplomacy's practitioners will be called upon to be creative and innovative actors in an effort to inspire people to "turn to Earth" as a model for conduct and long-term sustainability.[6] Ultimately, Sustainable Diplomats will be among the most articulate advocates for bringing religious and ecological analysis of international affairs into the core practices of diplomacy.

In a world of escalating economic competition, the language employed to describe the search for increasingly scarce resources is more and more evident in public discourse. Yet the language of this book comes not from the vocabulary of the marketplace, which in the Realist approach makes a commodity of all nonhuman (and much human) life. Rather, this project is concerned with the language that reflects people's spiritual ties, their subsequent understanding of religious responsibility to their greater surroundings, and their concern for future generations. It is from this language that new policies will emerge, which will meet people where they live while inviting them to seek new directions for the future. Such a language emerges from the awareness of living within a greater biosphere that is not designed by human hands. This language is heard in conversations that often begin in cafés, at schools, at dinner tables, and in many different places of work rather than in the traditional corridors of power.

The Case Study

The choice of Spanish–Moroccan relations as the geographic focus of this book is based on four considerations. First, the modern relationship between Morocco and Spain brings into high relief the stark differences between two countries whose economies, political affiliations, government structures, and traditions differ. Second, the Moroccan–Spanish relationship offers a wonderful opportunity to explore how a religious analysis of international relations might serve to interpret relations between what are two increasingly contentious political entities: those nations in conflict whose respective religious cultures are defined by Islam and Christianity. Third, this choice of case study reflects my own long-standing fascination with the study and practice of Muslim–Christian dialogue, and my abiding interest in exploring the ecological dimension of interreligious dialogue within international relations. Finally, despite any and all differences between Morocco and Spain, it is the contention of this inquiry that together, through a common history, genealogy, and geography, Spain and Morocco comprise a single bioregion. The acknowledgment of this common identity could be the basis of an entirely new relationship between these two countries.

My desire to explore the relationship between Spain and Morocco is also drawn from personal experience. During two years of working as an English teacher in Seville, in the southern Spanish province of Andalucia, I spent nearly all of my free time traveling in Morocco. The more time I spent crossing the Moroccan–Spanish border, the more symmetry I came to see between the people, land, language, culture, and architecture of these two nations. Through hearing and learning to speak the heavily Arabic-influenced Andaluz dialect of Spanish spoken in Seville, to working in the shadow of the twelfth-century minaret that is the symbol of the city, I began to understand how dependent southern Spanish culture is on its Muslim roots. Likewise, in having the privilege of living in Fez with a Moroccan family while pursuing Arabic studies, my eyes were opened to a deeper level of a rich and fascinating culture. Finally, by spending time with the Spanish-speaking Riffian farmers of Morocco's northernmost mountain range, and noting over and over the fact that so many northern Moroccans physically and culturally resemble their "European" neighbors across the Strait, it became less and less clear to me what the border that divides Spain and Morocco actually represents.

In Spain and Morocco, one encounters a fascinating history of two nations that have, over the centuries, taken the opportunity to invade and occupy one another's land and people. This rich common history is the backdrop for the modern conflict, which is the Moroccan–Spanish frontier: the principal setting of one of the most contested migrations of human beings seeking to cross (both legally and illegally) the North–South split. I believe that this setting, because of the intertwined history and cultural interdependence of the Spanish and Moroccan people, provides one of the best contexts to reflect upon how Muslim–Christian relations can be most constructively negotiated in the future.

Sources/Methodology

The methodology of this book will be based on comparative textual analyses of both the first-person words of people "on the ground" and selections from the Qur'an and the Bible. These texts will be used to fashion an ethical framework for a subsequent political analysis of three points of conflict between Morocco and Spain: conflicts over land, natural resources, and immigration. Additional scholarly texts and press articles are used, analyzed, and critiqued throughout this book. Among the texts cited will be those drawn from information gathered during 160 hour-long interviews conducted between June and October of 2000 with Spaniards and Moroccans, from Barcelona in northern Spain to Guelmim, near the edge of the Western Sahara. The use of interviews in this book will be guided by the qualitative anthropological approach, which acknowledges that the opinions and observations of individuals can never be seen as normative for an entire group of people. The use of interviews fulfills two goals of this project: (1) to offer the reader an opportunity to hear a variety of voices from across the Spanish and Moroccan social spectrums; and (2) to insure the inclusion of voices that are often excluded in the construction of political policy.[7] These interviews, based on a set of questions that invited the interviewee to describe his or her Ecological Location, captured portraits of farmers, fishermen, artists, students, construction workers, teachers, home makers, religious leaders, and the unemployed, among others. While no set of interviews (or textual analyses) can hope to paint the definitive and/or objective portrait of the highly complex Moroccan–Spanish relationship, I believe that this approach has provided a viable strategy for raising and clarifying some important questions, as well as guiding further study.

Beginning with an overview of one of the principal modern sources of the realpolitik analysis of international relations, this project will move on to engage the following sources: (1) a consideration of the work of some of the major proponents of a religious analysis of international affairs; (2) a synopsis drawn from historical texts recounting the evolution of the relationship between Spain and Morocco; (3) a comparative treatment of some Muslim and Christian scriptural narratives that engage the relationship between human beings and the land, natural resources, and those considered to be "other"; (4) selections from 160 interviews conducted with Spaniards and Moroccans describing their own personal Ecological Locations; (5) an analysis of a variety of books and articles detailing some of the theological, economic, political, and cultural influences that serve to help describe Morocco's and Spain's Ecological Locations as nation-states; (6) a treatment of the friction caused by Spain's continued holding of land on the Moroccan landmass; (7) the struggle over control of fishing grounds between Spain and Morocco; (8) a consideration of the phenomenon of the Moroccan–Spanish frontier as the setting for the majority of legal and illegal North African immigration into the nations of the European Union (EU); and (9) a survey and critique of the future applications of Ecological Realism and Sustainable Diplomacy, including a treatment of the future of Spanish–Moroccan relations in the context of the EU, the relationship between Muslims and Christians, the growing tensions and opportunities presented by the North–South split, and the importance of coming to view Spain and Morocco as one bioregion.

Making New Connections

This work is an attempt to contribute to the building bridges among the disciplinary perspectives of the fields of Christian and Islamic Ethics, International Relations, History, Anthropology, and Environmental Studies. This project emerges from the conviction that all these fields must be placed into more comprehensive dialogue with one another.

As a Christian ethicist, I have written this text from a particular disciplinary perspective. As a student of diplomacy, Islam, and environmental ethics, I have undertaken in these pages to challenge religious exclusivism, and to invite ethicists from across the religious spectrum into dialogue with a variety of disciplines that are all too often kept separate. While the religious context of this work is Muslim–Christian dialogue, the foundational arguments put forward in this text could apply to many other dialogues among many other faith traditions. Advocating such work from the perspective of Christian Ethics is therefore an invitation to broaden a number of conversations including those within the Christian community itself.

With noteworthy exceptions, the discipline of Christian Ethics is not generally inclined to engage the subject of International Relations comprehensively. Most Christian ethicists approach political ethics from an individual, small group, or domestic context.[8] It is true that the religious implications of political exchanges and the political qualities of interreligious dialogue have been the focus of the work of more than one Christian ethicist. However, when political analysis reaches the level of examining relations among nation-states, most Christian ethicists have traditionally ceded their ground to secular scholars. While Reinhold Niebuhr's "Christian Realism" is one of the more noteworthy exceptions to this rule, its propensity to validate a Realist analysis (along with its admitted attending moral analysis of the Realist world) did little to distinguish Christian ethics from its secular counterparts in political theory.

In contrast to the current approaches with which Christian Ethics has taken to engaging international relations analysis, the approach of this project is a constructive, supplementary, and revisionary challenging of the currently accepted norms. To begin with, this project work is an effort to push beyond the limits of the classical Realist paradigm of International Relations in a manner that encourages others in many fields, including Christian and Islamic Ethics, to question the rationale of Realism's primacy. Second, by modeling an alternative methodology to the dominant approach of foreign affairs analysis, this book, taken as a whole, is an attempt to make persuasive the claim that a religious analysis should be central to the discipline of examining relations among nation-states. In addition, by adapting Daniel Spencer's Ecological Location and Wackernagel and Rees' Ecological Footprint to the task of describing the theo-ethical norms of a particular nation-state, the following pages are an effort to commend the importance of the ecological dimension of religious and political analysis to audiences not yet informed by this perspective. Finally, by modeling a new religious approach to international affairs analysis, through Ecological Realism and Sustainable Diplomacy, it is hoped that this book can be a source of inspiration to those in the field to recognize foreign affairs as a legitimate and critical area for future work in all traditions of Religious Ethics.

A Synopsis of the Chapters

This book is divided into six chapters, whose content and objectives will be presented in the following order.

Chapter 1, "Interpreting Human Communities in Conflict," will present and critique some of the core precepts that guide a classical Realist analysis of international relations, by focusing on the work of Hans Morgenthau, alongside those who have argued the importance of considering religion in the analysis of relations among nation-states. This chapter will then turn to introducing the concepts of Ecological Location, the Ecological Footprint, Ecological Realism, and Sustainable Diplomacy as tools to employ in the analysis of international relations. Chapter 1 will conclude by naming the normative guideposts or framework of Sustainable Diplomacy: solidarity, participation, sufficiency, equity, accountability, material simplicity, spiritual richness, responsibility, and subsidiarity.[9]

Chapter 2, "The Foundations of the Eco-Historical Landscape of Moroccan–Spanish Relations" will present a historical overview of Moroccan–Spanish relations through the lens of Ecological Location. The chapter will conclude by presenting the three sources of conflict between Spain and Morocco that will be the subject of the following chapters: the conflict over land (Ceuta and Melilla), the conflict over natural resources (fishing rights), and the conflict over people (immigration, legal, and illegal).

Chapters 3–5 will follow the same structure. Each chapter will begin with the words of one Spaniard and one Moroccan on the theme of the chapter. Passages from these interviews will be followed by a scriptural comparison of the Bible and the Qur'an on a theme that speaks directly to the topic of the chapter. In light of the statements of the interviewees and scriptural selections, common and contrasting ethical principles will be proposed and analyzed. Each chapter will then present a portion of the Ecological Footprint illuminating an area of Spanish and Moroccan consumption. The chapters will end by returning to the ethical principles first presented, using them as a framework for a political analysis of the point of conflict being considered.

Chapter 3, "The Conflict Over Land: The First Human, Land Use and the Two Cities," will begin with the words of one Moroccan and one Spaniard on the subject of the Ceuta and Melilla, Spain's two city-states on the Moroccan land mass, and will follow with a comparison of the Biblical and Qur'anic narratives on the creation of Adam. These two elements will inform the construction of guiding ethical principles for analysis. Chapter 3 will then present an Ecological Footprint analysis of Spain and Morocco on the topic of land consumption. This chapter will then turn to a political analysis of the conflict over Spain's two holdings on the Moroccan landmass, using the framework of the ethical principles established at the beginning of the chapter.

Chapter 4, "The Conflict Over Natural Resources: The Tree of Life and the Tree of Being, the Consumption of Natural Resources, and the Fish Wars," will begin with the words of one Spanish and one Moroccan interviewee speaking on the subject of resource consumption and scarcity. A comparison of the Christian Tree of Life and the Islamic Tree of Being and the common and contrasting ethical principles drawn from these sources will follow. The chapter will then turn to the portion of the Ecological Footprint that addresses ocean resource consumption. Finally, by using the common and contrasting ethical principles proposed at the beginning of the chapter, chapter 4 will offer a political analysis of the ongoing conflict between Spain and Morocco over fishing rights.

Chapter 5, "The Conflict Over People—The Story of Abraham and Ibrahim and the Strangers, the Consumption of Illegal Human Labor, and the Conflict Over Immigration," will begin with some interviewees' reflections on immigration, followed by a comparison of Biblical and Qur'anic treatments of the story of Abraham/Ibrahim and "the strangers." After offering common and contrasting ethical principles in light of the words of the interviewees and the scriptural accounts, the chapter will then propose expanding the Ecological Footprint to include the impact and cost of illegal labor. In its conclusion, chapter 5 will use the chapter's earlier established ethical principles as a framework for a political analysis of the ongoing conflict between Spain and Morocco over immigration.

Finally, chapter 6, "The Future of Sustainable Diplomacy" will examine the possible future roles of the Spanish–Moroccan relationship, looking at the two countries as a common bioregion, as a point of contact between the EU and the Maghreb, and as a relationship between a predominantly Christian and a predominantly Muslim country. The chapter will then turn to the goals of Sustainable Diplomacy, and the impediments to implementing it as a practice. These subjects will be followed by a treatment of the "cross-traditional sins" that plague the followers of both Christianity and Islam, as well as some general common ethical principles. The conclusion of chapter 6 will pose some of the questions that remain in light of this study, and challenge Muslims and Christians to defy traditional calls of religious exclusivism and other forms of separation, and to make a pilgrimage together.

As I have noted, this book will engage several fields: Christian and Islamic Ethics, International Relations, History, Anthropology, and Environmental Studies. Although I am neither a Spaniard, nor a Moroccan, I draw upon these fields as a committed American student who wants to understand the subject at hand on a more profound level. The reader of this text is therefore invited to join me in this inquiry, bringing all your questions, along with an openness to engage in new conversations. With this in mind, I offer this writing as simply one particular interpreter of ideas, who desires many further conversations and many more conversation partners.

CHAPTER 1

INTERPRETING HUMAN COMMUNITIES IN CONFLICT

There is a mountain between us
And the roads are closed
And the messengers are few

Some people dig springs
But jealousy prevents water from running
There is a mountain between us

I don't need to send any messengers to my God
There is a mountain between us

I am standing in front of His gate
There is a mountain between us

My friend is as rare as a tajine
There is a mountain between us

He is never sated the person who eats it
There is a mountain between us

I walk and you walk
And God's will brought us together
There is a mountain between us

Who came first
And who came last,
Nobody can tell me
There is a mountain between us[1]

—*from a Tamazight song by Ali Ouzineb*
and Mohamed Qat

I nterpreting the dynamics of human communities in conflict presents a variety of challenges. Historians search for the root cause of a dispute in order to seek modern solutions. Economists struggle with the ins and outs of trade and the disparate value of goods and services in the name of striking an understanding. In turn, diplomats grapple with the work of the historian and the economist as they seek a stabilizing balance of power between warring factions or nation-states. All three of these forms of analysis make significant contributions to the work of conflict resolution, and each provides tools and insights lacking in the other two.

"There is a mountain between us," sings the writer of the song. It is a mountain that both separates and connects, while helping to form the identities of those who live on either side of it. Those who interpret human conflict are obliged to identify both the barriers and pathways to cooperation. "The roads are closed," sings the writer of the song, "and the messengers are few." For the author of these words, the land itself reflects not only the actions but also the temperament of the people. The hard and substantive work of digging wells is negated by human jealousy, which prevents the water from running. In turn messengers, like water itself, are in short supply, perhaps due to the same factors that have blocked the springs. The land is thus a connector, a barrier, and a reflection of human failings. Despite all these variables, however, the Divine appears to be constant. "I don't need to send any messengers to my God. . . . I am standing in front of His gate," sings the writer. God is present and access to God is not contingent on the quality of any road. "I walk and you walk, and God brought us together." The mountain remains, but its ability to separate is laid aside. God's transcendence allows humans to move beyond what appear to be insurmountable barriers. And what were formerly seen as barriers now serve a different role.

> Who came first
> And who came last,
> Nobody can tell me
> There is a mountain between us[2]

The human relationship to land and to the Divine are thus seen as the means to building human relationships, rather than the source of acrimony and conflict.

What is missing in the prevailing analyses of relations among nation-states? What are the questions not being asked by the dominant schools of thought in seeking conflict resolution? Could the land and the greater biosphere play a role in promoting understanding between different nation-states? Is the role of religion potentially positive in the work of conflict resolution?

This chapter is an attempt to go beyond the boundaries of the prevailing schools of international affairs by raising the issues of power, language, religion, sustainability, and transnational cooperation in a new light, informed as much by the concerns of international relations theorists as it is by theologians and farmers. The degree to which such a project succeeds is based on its ability to find a common language that crosses the boundaries of varying disciplines and ideologies. The time has arrived to examine the likelihood that the common language is found in the ecosphere itself.

Statesmen, stateswomen, and those who wage war use maps, or other attempts to express the physical contours of the Earth, as they begin to consider their options. They most often do so in an anthropocentric light. The aim of this writing is not only to question the logic of anthropocentrism, but to explore the implications of considering the ecosphere itself as an actor on the international stage. This chapter will argue that the ecosphere itself should not be viewed as a neutral actor; instead it should be seen as an active source of common language, a focal point in the sharing of stories of faith, and the ground for learning about a new way for humans to live sustainably together with nonhuman creation. It has been said that the biosphere itself holds the key to lasting human cooperation. Humans, who are embedded within its confines, are ultimately obliged to confront the limitations of their current practices of consumption and pollution, or risk their own demise.

Clearly, there is a need for a new and fresh approach to the conduct of diplomacy, not simply as an academic exercise but as a necessity. The time has arrived to move beyond the old worldviews that have guided our definitions of power and the singular nation-state. Now we are called to ask ourselves an entirely new set of questions, ones that include the concerns of consumption, language, religion, race, ethnicity, culture, nationality, and bioregions. We must ask ourselves what will be required to foment cooperation, not only between leaders of nation-states, but also between entire transnational populations, across the contentious boundaries of race, religion, ethnicity, nationality, class, and gender.

The models of diplomacy that have focused solely on singular state-to-state relations must now give way to a more organic model, which is influenced not by artificially designed borders, but by the dictates of the biosphere. At the same time the human propensity to separate the human population from the biosphere or to hierarchically stratify nonhuman creation must be rejected in favor of an understanding that humanity is deeply and inextricably embedded in the entire ecosphere. Such a human relationship with the biosphere is dialectical, with each side influencing the other. Further, such a relationship overrides any human claims of ownership or control over human or nonhuman creation. Finally, an understanding of the ecosphere as ultimately the central player in human relations turns on its head our previously understood definitions of power, including its origin, form, focus, and future.

This chapter begins by considering one of the classic approaches of international relations analysis, the Realist school, followed by the introduction of three newer methodologies. One approach focuses on the role of religion in conflict resolution, as a corrective to traditional ways of understanding relations among nation-states. The second methodology centers on the Ecological Locations of the nation-states in dispute, as potential sources of identity, understanding sustainability, and cooperation. The third approach seeks to determine the Ecological Footprint a nation-state makes upon the Earth in its consumption, destruction, and preservation of natural resources.

"There is a mountain between us," sings the author of the song. What role can this mountain play? Where do the people understand the mountain to have come from? Who owns the mountain and what distinguishes the people living on one side of the mountain from the other? What are the understandings of the role of the Divine in this relationship? The following pages are an effort to approach old dilemmas with new questions, in the hope of gaining a new perspective.

Political Realism

While the Realist school is hardly the only school seeking to analyze relations between nation-states, it can in many ways be viewed as a "classic approach."[3] By examining Realism, we are given a window into diplomacy's formative past and sometime present. Thus, in order to appreciate how far we still have to move, an analysis of Realism is one way of understanding many of the long-ingrained habits in the current practice of foreign policy.

In the Realist view, diplomacy is governed by a broad set of secular and material assumptions regarding human exchanges, and understands nation-state conduct as guided by the desire to acquire, retain, and project power. Scholars of foreign affairs trace the roots of the Realist school to the Enlightenment, and they attribute its understanding of human conduct to a lineage of influences that include Machiavelli, Auguste Compte, and Max Weber.[4] In this regard, Realism is a synthesis of the work of many thinkers. Hence, the Realist school's materialist approach to international affairs is an attempt to anchor the discipline in readily identifiable and previously acknowledged forms of analysis.

One of the more prominent twentieth-century proponents of the Realist school was Hans J. Morgenthau. While Morgenthau was only one of Realism's many proponents and definers, he offers a familiar and useful approach to defining Realism. Much of Morgenthau's work was an attempt to distill what we can call "classic Realism." In his book, *Politics Among Nations*, Morgenthau outlines what he describes as the six principles of political Realism:[5]

1. Political realism believes that politics, like society in general, is governed by objective laws that have their roots in human nature. In order to improve society it is first necessary to understand the laws by which society lives. The operation of these laws being impervious to our preferences, men will challenge them only at the risk of failure.[6]

For Realism, the objective laws that govern society were long established in antiquity. One law, according to Morgenthau, is that individuals and nation-states are predisposed to act in their own self-interest. The task of Realism is thus to "distinguish in politics between truth and opinion."[7] In Realism, the political objectives of a nation-state are therefore best ascertained by observing its actions, not through listening to the rhetoric of its leaders. Such has been the case throughout history. Drawing conclusions regarding observable nation-state conduct is therefore, in the eyes of the classic Realist, a practice that is rational and objective. For Morgenthau, a theory's veracity is enhanced rather than diminished by its longevity.[8]

Morgenthau's second principle goes to the heart of his definition of Realism:

2. The main signpost that helps political realism to find its way through the landscape of international politics is the concept of interest defined in terms of power.[9]

For the classic Realist, heads of state are guided by the desire to and necessity of acquiring, retaining, and projecting power. In Realism, interest defined as power is the most apt means of imposing intellectual discipline and rational order on the field of international relations.[10] This is because by viewing relations among nation-states through the lens of interest defined as power, one can see more clearly the consistency in the actions of seemingly disparate countries. According to Morgenthau, this consistency remains "regardless of the different motives, preferences and intellectual and moral qualities of successive statesmen."[11] For this reason, classic Realism holds that it is not important to understand the motives of a statesman or stateswoman but rather his or her "intellectual ability to comprehend the essentials of foreign policy."[12] Classic Realism, does not disregard the impact of political ideals or moral principles, it requires "a sharp distinction" between what Morgenthau calls "the desirable and the possible."[13]

Morgenthau's third principle of political Realism reads as follows:

3. Realism assumes that its key concept of interest defined as power is an objective category which is universally valid, but it does not endow that concept with a meaning that is fixed once and for all.[14]

Therefore, power can take on many forms and be "anything that establishes and maintains the control of man over man."[15] According to Morgenthau,

. . . power covers all social relationships . . . from physical violence to the most subtle psychological ties by which one mind controls another. Power covers the domination of man by man, both when it is disciplined by moral ends and controlled by constitutional safeguards . . . and when it is that untamed and barbaric force which finds its laws in nothing but its own strength and its sole justification in its aggrandizement.[16]

According to Morgenthau, despite its potential for destruction, power can temporarily be brought into a system of checks and balances. Among the most useful of these systems is a balance of power. According to Morgenthau, a balance of power is "a perennial element of all pluralistic societies."[17] The concept that the potential for striking a balance of power exists across the spectrum of circumstance appeals to the Realist viewpoint, which seeks to draw on what it sees as an extant phenomenon rather than to impose what Morgenthau would call an "abstract ideal."[18] This is because according to the dictates of classical Realism, transformation only occurs when leaders of nation-states manipulate such "perennial forces" as a balance of power in the search for peace and security.[19]

Morgenthau's fourth principle of Realism addresses the topic of political ethics:

4. Political realism is aware of the moral significance of political action. It is also aware of the ineluctable tension between the moral command and the requirements of successful political action.[20]

Realism does not allow for the application of abstract, universal ethics to the conduct of nation-states. Rather, moral principles must be interpreted "through the concrete circumstances of time and place."[21] Given this standard, the classic Realist argues that while the nation-state "must judge political action by universal moral principles, . . . the moral principle of national survival" is paramount in the conduct of political action.[22] For this reason, notes Morgenthau,

> Ethics in the abstract judges action by its conformity with the moral law; political ethics judges action by its political consequences.[23]

Interest defined as power provides a universal principle for conducting political exchanges and the best means of anticipating future nation-state conduct. At the same time "prudence—the weighing of the consequences of alternative political action," is considered to be "the supreme virtue in politics."[24]

Morgenthau's fifth principle of Realism reads as follows:

> 5. Political realism refuses to identify the moral aspirations of a particular nation with the moral laws that govern the universe.[25]

For classic Realists, political Realism dictates that nations must resist the temptation to see themselves as the vanguard of what is good and true. For while it is possible to be familiar with universal moral laws, it is quite another thing to know "what is good and evil in the relations among nations."[26] For this reason, interest defined in terms of power is once again an essential concept, for interest defined as power allows one to judge the behavior of all nations, including one's own, by the standards of self-interest. Thus, each nation is able to form policies taking into consideration the interests of other nations "while protecting and promoting those of [their] own."[27]

Finally, Morgenthau's sixth principle of political Realism stresses the preeminence of the political standard over other standards of analysis in international relations:

> 6. The political realist is not unaware of the existence and relevance of standards of thought other than political ones. As [a] political realist [one] cannot but subordinate these other standards to those of politics.[28]

Classic Realism stresses the autonomy of the political over other standards of thought in international relations. For example, according to Morgenthau, a strictly legalistic standard of international relations could undercut a nation-state's ability to protect its interest defined as power. Thus, an illegal seizing of territories by a hostile nation will not necessarily be reversed by a strictly legalistic response. That is not to say that other standards are of no consequence. The economic, legalistic, and moralistic standards, to name three, exert critical influence in the realm of international relations. For the classic Realist, however, these other standards are subordinate to the political standard of interest defined as power.[29]

Hans Morgenthau's version of the Realist school defines politics as it is still practiced by many on the international stage. Certainly this can be said of Morgenthau's insistence on defining interest in terms of power. Desiring to place political theory in the realm of the hard sciences, the Realist school emphasizes the universal over the exceptional. Likewise, classic Realism's law of self-interest and rejection of a moralistic–legalistic understanding of international relations underscores a desire to render a uniform portrait of relations among nation-states. At the same time, classic Realism insists that conclusions can only be based on incontrovertible evidence. Thus, a national leader's words are of less consequence than his or her actions.

What is the Realist school's ideal arrangement among countries? In a world where interest is defined in terms of power, classic Realism finds equilibrium in striking balances of power among competing nation-states. Such a balance, however, is temporal and requires constant maintenance on the part of national leaders who must work to distinguish between what is desirable and what is possible. What is impossible, notes Morgenthau, is the capacity to know what is good and evil in the realm of international relations.[30] Knowledge is in the doing. What is good and what will work will not emerge from a universal idea of good but rather through the concrete circumstances of practical experience.

Yet what is missing from classic Realism's portrait of international relations? To begin with, the Realist school is exclusively secularist–materialist in its assumptions regarding human motivations. For this reason, a Realist analysis does not privilege appeals to moral norms beyond the necessity of increasing nation-state prestige and power. According to the political theorist Stanton Burnett, the Realist school's arrival into international affairs theory was a response to a perceived lack of scientific theory in the field.[31]

> In its thirst to imitate the physical sciences (and to gain, therefore, the success and prestige the physical sciences have in our society), the Realist school, along with its offspring and principal competitors, was dogmatically, unflinchingly secular. Its denial of human—including religious and spiritual—factors was a mere part of its denial of all cultural factors as significant in the shaping of the behaviors of states (the only actors on display).[32]

At the same time, classical Realism's worldview is strictly anthropocentric. Power is defined in exclusively human terms, while moral principles find their origin in human choice and design. Realism argues that human efficacy lies in the recognition of the ultimacy of human power. Thus, for classic Realism, the amelioration of the problems the human community faces will be brought about by selective human control, governed by the knowledge and actions of nation-state leaders.

Realism's insistence on the primacy of the nation-state sets the stage for a variety of assumptions. In Realism's analysis, "success" is a compartmentalized commodity, relegated to the individual perceptions of each nation-state. When competing perceptions find harmony, conflict is reduced, though this is considered by the classic Realist to be a nonnatural state of being.[33] In addition, by insisting on the nation-state as the ultimate arbiter, classic Realism discounts other sources of influence and power, including NGOs and the greater biotic system itself. Finally, classic Realism rejects the possibility that the spiritual life of a nation-state's inhabitants and their leaders might bring to bear influence on the conduct of relations among nation-states. At the same time, classic Realism ignores the presence and importance of the current state of the biosphere, and the need to change human ecological conduct. As a post-facto form of analysis that focuses on human power, classic Realism is thus unable to provide models of sustainability.

Religion and International Relations

Classic Realism discounts the possibility that religion is a principal source of conflict and conflict resolution. Yet in the wake of the Cold War, from which the Realists drew much of their direction, different sources of conflict and negotiation have necessarily emerged. Douglas Johnston, in his essay entitled "Introduction: Beyond Power Politics," offers but a few:

> With the decline of the East-West confrontation and most of its regional manifestations, few of the conflicts that evolve will be rooted any longer in the old Cold War ideologies. Instead, most will derive from clashes of communal identity, whether on the basis of race, ethnicity, nationality, or religion. Such disputes tend to occur at the fault lines between rival nationalities or in situations where societies are suffering from the strains of economic competition and rising expectations. These are the most intractable sources of conflict, and they are the sources with which conventional diplomacy is least suited to deal.[34]

At the same time, Johnston reminds his readers that religion can be a binding force rather than a fractious one. Parties in conflict can be appealed to on the basis of faith, often with positive results.

> We also inadequately appreciate the transformational possibilities that exist when the parties involved in a conflict can be appealed to on the basis of shared spiritual convictions or values. Implicit in the latter is the prospect that, under the right conditions, the parties can operate at a higher level of trust than would otherwise be possible in the realm of realpolitik. This is not to suggest that it is an "either-or" proposition with regard to the spiritual and the secular. More likely, it is a "both-and" phenomenon in which a breakthrough at the spiritual level is made possible once the political, economic, and security "planets" have been brought into some kind of proximate alignment.[35]

Thus, the introduction of religion as an element of consideration is not introduced at the expense of traditional secular modes of analysis. Rather, religion illumines what has until now been seen as exclusively secular. Such a phenomenon is seen as a direct contradiction to those who view the secular and the spiritual as mutually exclusive. Edward Luttwak, in his essay "The Missing Dimension," explores the inherent tension in this claim.

> Astonishingly persistent, Enlightenment prejudice has remained amply manifest in the contemporary professional analysis of foreign affairs. Policymakers, diplomats, journalists, and scholars who are ready to overinterpret economic causality, who are apt to dissect social differentiations most finely, and who will minutely categorize political affiliations and are still in the habit of disregarding the role of religion, religious institutions, and religious motivations in explaining politics and conflict, and even in reporting their concrete modalities. Equally, the role of religious leaders, religious institutions, and religiously motivated lay figures in conflict resolution has also been disregarded—or treated as a marginal phenomenon and hardly worth noting.
>   One is therefore confronted with a learned repugnance to contend *intellectually* with all that is religion or belongs to it—a complex inhibition compounded out of the peculiar embarrassment that many feel when faced by explicit manifestations of serious religious sentiment; out of the mistaken Enlightenment prediction that the progress of knowledge and the influence of religion were mutually exclusive, making the latter a waning force; and sometimes out of a willful cynicism that illegitimately claims the virtue of realism.[36]

Clearly, the fear of the ostensibly unquantifiable has frightened analysts from the pursuit of religion as an analytical tool. The influence of religious faith and culture cannot be quantified in the same manner as an economic index. Such inadequacies, however, do not by any means limit religion's influence in the realm of the political.

The Cost of Neglecting the Religious Dimension of Analysis: Iran, Iraq, Afghanistan, and U.S. Foreign Policy

It is interesting to note that three of the greatest blunders in late twentieth-century Western foreign policy have been made by a country (the United States) whose ostensible religio-cultural identification is Christian, and whose errors were made in the course of attempting to engage three Islam-identified nation-states: Iran, Iraq, and Afghanistan. While retrospective consideration of these three cases clearly illustrates the utility of valuing the role of religion in nation-state identity formation, the original absence of such an approach is particularly ironic when considering the depth of analysis that was lavished on secular factors governing the conduct of these three countries and their respective populations.[37]

The case of U.S. policy formation with regard to *Iran* provides what is possibly the richest example of the consequences of neglecting the religious dimension of analysis. According to Edward Luttwak, American errors can be attributed to three specific and erroneous assumptions: (1) that opposition to the Shah was not based on a religious rejection of Westernizing modernization and the authoritarian regime that promoted it; (2) that only "modern" Iranians (i.e. secular) were likely to have the means and motivation ultimately to control Iran's fate; and (3) that a religiously identified movement, such as the one whose leader (Khomeini) was living in exile in Paris, could never command the broad base of support required to govern a nation of the size and diversity of Iran.[38] Barry Rubin, a political analyst of Iranian politics, argues that these three false assumptions prevented the United States from being able to anticipate, or even to understand, many of the choices that the new Iranian leadership ultimately made, because those possibilities did not reside within the boundaries of what was then considered to be "rational" behavior. These choices included Tehran's seizing of the American Embassy, its choice to continue a war with Iraq "long after the battle was counterproductive," and Khomeini's issuing of a fatwa calling for the murder of Salman Rushdie at the very "moment Iran [desperately] needed Western investment for reconstruction."[39] Clearly, in a secularist context, many of the Iranian leadership's choices have defied the rules guiding the market economy or a classical Realist analysis of nation-state conduct. Yet as an expression of policy based on the privileging of religiously guided motivations, all three of these events can be understood, and some, perhaps, might even have been possible to anticipate.

As the case of *Iraq* continues to unfold, U.S. policy formation remains consistent in its reluctance to privilege the religious influences that could serve to contextualize much of Iraq's behavior. Backed by the United States early in his leadership of Iraq, Saddam Hussein and his Ba'ath party were seen as an important Sunni bulwark in the overall U.S. strategy to contain what was viewed as the offensive geographic aspirations of the Shi'i leadership of Iran. Having equated Saddam Hussein's nominal identification with the Sunni branch of Islam with political moderation, the United States and much of the West was taken by surprise when Iraqi troops seized neighboring Kuwait. The subsequent coalition building and Gulf War that followed marked yet another lost learning opportunity for the West. Having decimated Iraqi forces, the coalition army stopped short of taking Baghdad—at the time, a choice that was the subject of much speculation. To enter Baghdad, however, would have risked bringing about the end of what was still considered to be the preferable leadership of the Sunni Iraqi minority over that of the majority Shi'i population. Eliminating Saddam would have also risked the fracture of Iraq along ethnic as well as religious lines, not only in terms of the Shi'i majority, but also because of the sizable Kurdish population within the boundaries of Iraq who harbor their own aspirations for independence. The ultimate American choice, to maintain the status quo in Iraq by employing a modified policy of containment, was clearly a decision guided by the assumptions of a classic Realist analysis and a privileging of the will of market forces. An Iraq that is broken into pieces along the lines of Shi'i, Sunni, and Kurdish populations would seriously undermine the U.S. goal of using Iraq as a buffer state in the containment of Iran. From a Western perspective, the breakup of present-day Iraq would also greatly complicate future petroleum extraction and export from the region. Making the religious and ethnic borders within Iraq national borders would require multilateral negotiations, whereas the current configuration requires only bilateral exchanges. Such policy decisions on the part of the United States and its allies, however, have not changed the fact that Iraq remains highly unstable; a situation that brings into question its capacity to be perpetually governed by one central government due to powerful religious, ethnic, and secular schisms within the population. Yet despite these factors, U.S. policy formation remains clearly guided by a belief that privileging a classic Realist analysis is the best means of obtaining future security in the region, an assumption that defies most of the lessons of recent history. Regretfully, this same Realist worldview has been sustained in George W. Bush's most recent military incursion into Iraq.

The American experience in *Afghanistan* is yet another example of a secular Realist analysis diminishing the possibility of successfully anticipating the trajectory of policy choices involving religious actors. In the wake of the 1979 Soviet invasion of Afghanistan, American policy makers explored a number of options, finally concluding under the Reagan administration that the backing of Muslim Afghani insurgents employing guerilla tactics would be the most cost-effective approach with the best chance of expelling the Soviets. In this case, U.S. choices were guided by more than a teleological approach, which saw the guerilla forces as the most effective extant option toward realizing the goals of a purely secular analysis. By diminishing and universalizing the importance and specific nature of the religious ideals guiding the Afghani fighters (who, unlike their American backers, did not see their struggle in the context of the ongoing Cold War), U.S. analysts were unable to anticipate the link of governance and religiously driven Civil War that would emerge from the vacuum created by a Soviet pullout. Unfortunately, this willful blindness to such "facts on the ground" persists throughout the conduct of American foreign policy in Afghanistan in a post–9/11 world.

Ultimately, the secularist assumptions made by the United States and many of its Western allies are among one of the greatest shortcomings of the classic Realist approach: principal among these is the assertion that in the modern era, the influence of religion must be diminished in order for a nation to form a national self-consciousness. At the same time, when some secular analysts do express a willingness to take seriously the power and influence of religious forces, they most often do so by treating each religion as though its followers were an ideological and/or theological monoculture, without nuance, internal debate, or conflict. Thus, Western acknowledgment that there are different types of Islam, let alone different types of Sunnis and Shi'is, is an admission that is most often only reluctantly made after the failure of a policy.

Approaching every religiously identified political actor/movement as a distinct and unique entity, not only in terms of religious allegiance, but also in ethnicity, race, class, and specific geographic location, should clearly be the foundation of any successful prescriptive approach to forming the foreign policies of the future. As Stanton Burnett has observed:

> . . . the fact that modern sociology (and its offspring, modern political science) lacks tools of analysis for important parts of the life of [people] and nations should not impede sober consideration and serious "use" by diplomats of this whole range of ethical, religious, spiritual, philosophical, and mythical phenomena, which are important to political actors and potentially helpful interveners.[40]

Yet the conduct of foreign policy is hardly limited to the guidance and control of conflict. At its best, diplomacy's business is the art of building peaceful and mutually beneficial alliances. Building any type of lasting and cooperative peace between two or more nations requires the mutual acknowledgment and understanding of each nation's culture and the aspirations harbored by its people. Such goals, it would therefore seem, cannot be attained in the absence of analyzing the impact of religious movements and culture upon the political history and present-day circumstances of nation-states.

Deepening the Reading of a Nation-State: Ecological Location

Understanding the underpinnings of conflict between nation-states often requires the use of unconventional means. As we have seen, international relations analysis that dismisses a religious analysis risks misunderstanding the conflict at hand. Yet just as religion has often been neglected in the course of analyzing relations among nation-states, so too has the role that nonhuman nature plays in the lives of the nations and the people who live within them. In many respects these two schools of analysis, religious and ecological, are deeply linked. One often serves to instruct us on how to better understand the other.

The Christian Ethicist Daniel Spencer proposed a means of examining the relationship that the biotic and non-biotic portions of creation interact, influence, and inform one another. He calls his approach *Ecological Location*.

> By *ecological location*, I mean enlarging the term *social location* to include both where human beings are located within human society and within the broader biotic community, as well as conceiving other members of the biotic community and the biotic community itself as locatable active agents that historically interact and shape the other members of the ecological community, including

human beings. Just as social location is an anthropocentric term that helps us to pay attention to how human identities are multiply formed with respect to various lines of human difference, ecological location is an ecocentric term that recognizes that human epistemologies—how we see and interpret the world—are also shaped by our relationship with the land and other creatures in our broader biotic environment.[41]

Ecological Location represents not only a new way of seeing but a new way of interpreting the human relationship to greater Creation. At the same time, Ecological Location forces us to admit that nonhuman members of the biotic community are themselves active agents that shape the lives of human communities. In one respect this can be seen in terms of power. As Spencer has noted:

> In recent years, ethicists have highlighted the importance of paying attention to the social locations of persons and communities as a way to reveal power relations in society. Yet as my own upbringing illustrates, it is not only social patterns, but also biophysical and ecological ones that influence how we act and see in the world. If, as Lewis Mumford suggests "all thinking worthy of the name must now be ecological," combining liberationist and ecological approaches suggests the need for an expanded concept of ecological location.[42]

Ecological Location stands at the crossroads between our concern for human perception and the nonhuman players that shape it. Ecological Location goes beyond the term "social location" and forces the viewer to concede that the environment itself shapes our worldviews. For too long now we have held the human experience at the center while regarding non-human nature to be inert. Within the context of Ecological Location, social location becomes part of a broader web, where "human and non-human creatures and communities are situated with respect to other members of the biotic community."[43] In this respect, Ecological Location "is the relevant whole or context that must be taken into consideration in ethical reflection."[44]

> The concept of ecological location is a logical outcome of liberationist efforts to deconstruct the nature/culture dichotomy that renders nature inert and invisible with respect to human affairs. Instead of the human social realm being seen as the only valued context, it must be understood as part of the broader ecological web of relations. Hence one's social location is a distinctive but interconnected part of a larger ecological location: where human and non-human creatures and communities are situated with respect to other members of the biotic community. How we are shaped to see and act in the world results from a complex interplay of physiological, social, cultural, and *environmental/ecological* factors. For ecological ethics (and, I would argue, the vast majority of social ethics), ecological location is the "relevant whole" or context that must be taken into consideration in ethical reflection.[45]

Thus, Ecological Location radically expands the scope of human formation, while at the same time emphasizes the centrality nonhuman nature.

Determining the Ecological Location of individuals or groups requires that we become familiar with their Creation stories, which are an important source of "social and ecological blueprints for how [people] organize social relations and interactions."[46] Creation stories give voice to what is often otherwise left silent in the human/nonhuman relationship. Creation stories are a source of history. Though they are not always literal, they are nearly always powerful touchstones through which human communities attempt to come to terms with their relationships, good or bad, with greater Creation. For this reason, Ecological Location recognizes "the spiritual dimension of human interactions and histories with particular places, habitats, and geographies."[47]

> . . . ecological location can help us to recover for ethical analysis what many indigenous religions have long recognized: the spiritual dimensions of the humanity/nature relation that develop for many and consciously or unconsciously influence the ways we see and act in the world. At least initially, human beings develop attachments not with nature or the biosphere in the abstract or universalized sense, but rather with particular places, particular communities of animals, plants, bodies of water, weather patterns, rock formations, seasonal rhythms. Many believe there is little chance for ecological ethics to succeed in helping to reverse the ecological crisis without human beings developing a renewed sense of spiritual connection to the land and all its creatures and parts. Ecological location can be one part of building up what Mitchell Thomas has called an "ecological identity"[48] by drawing attention to the particularity of our spiritual and aesthetic relations with nature, and how this affects our way of acting and seeing in the world.[49]

The spiritual relationships that emerge within our respective Ecological Locations affect our relationships with both human and nonhuman nature. Ecological Location's focus on the human relationship with greater Creation obliges us to take a hard look at the spiritual histories of the people and land we are trying to understand. Ecological locations are particularized along the lines of context, time, and ongoing relationships. At the same time, Ecological Location is an invitation to examine the power relationships that exist among humans, and between humans and nonhuman nature.

> Ecological location can help us recognize simultaneously power differences within the human community and in humanity/nature relations. Attention to power in ecological location can show how both the dynamics of intrahuman community and the human/nature biotic community are built on either cooperation or domination—that is, relations that either sustain or deplete us socially and ecologically. Thus it can help to better integrate understandings of ecological and social justice as right relation, a critical component of a liberationist ecological ethics. Keeping the intrahuman dynamic of social location is an integral part of ecological location. It retains the concern of

liberationist thinking about how human social differences are constructed into social relations of domination and exclusion. Expanding this to ecological location shows both how these human social relations affect the wider biotic community as well as how they are affected *by* the limits and makeup of the biotic community.[50]

The themes of cooperation and domination are critical to a thorough reading of Ecological Location. Until now, the locus of domination or cooperation has been seen as emanating exclusively from the human side of the human/nonhuman relationship. Ecological Location expands this notion to include the power that nonhuman nature has to cooperate with or displace human endeavors. At the same time, Ecological Location calls attention to the fact that particular human groups must attribute their own power to their relationship with nonhuman nature and its limits.

Ecological Location requires that we understand history as ecological. Much of the current status of any group can be explained in terms of its historical propensity to cooperate with or dominate the broader biotic community. For this reason, "good ecohistorical analysis" according to Spencer "can help us to better understand the (often contradictory) mix of attitudes, values and practices that shape our current ecological locations."[51]

Finally, to even begin to understand a community's Ecological Location, one is obliged to go beyond simple familiarity with the tastes and preferences of its national leaders. One is instead obliged to attempt to become familiar with the lives of ordinary people, across lines of gender, ethnicity, race, faith, language, and geography. By trying to understand how Ecological Location plays out on the ground, Spencer's proposal provides a means to move beyond the limitations of assuming that a treaty signed by a handful of people will assure peace among millions.

Expanding the Scope: Ecological Location and the Ecological Footprint Analysis

Daniel Spencer's proposal of Ecological Location provides a powerful tool for analyzing the lives and circumstances of human and nonhuman individuals and communities. However, to analyze the Ecological Location of an entire country requires other clarifying tools for describing relationships and conflicts that prevail on the scale of the nation-state. One possible means may be found in Mathis Wackernagel and William Rees' notion of the *Ecological Footprint*.

Wackernagel and Rees have devised a way to measure the ecological impact an individual country makes upon the Earth. The authors first ask us to imagine that the city or town that we live in is covered by a large glass dome, within which we must meet all our food and energy needs and dispose off all our pollution by-products. Clearly, it would only be a matter of days before many cities' inhabitants would perish under such circumstances. This is because the majority of villages, cities, and nation-states rely on land and resources outside their immediate borders in order to survive and, in some cases, flourish. Environmental Footprint analysis thus invites its practitioners to calculate how large such an actual dome would have to be in order to sustain the human and nonhuman habitants of a particular region.

It is quite extraordinary to find how much larger many nations' footprints are than their actual borders. For example, Wackernagel and Rees estimate that if the entire Earth's population were able to adopt the level of consumption and pollution of the average American or Canadian, it would require three planet Earths to support all the inhabitants of our one planet.[52] The Ecological Footprint model provides an account of the flows of resources and pollution across national borders in an effort to educate its users as to their level of resource dependency and the degree to which some nations borrow or steal the sustainability of others. At the same time, the Ecological Footprint analysis is a way for a city or a nation-state to assess in cold hard numbers the challenge of making the regions they inhabit sustainable.

Like Spencer, Wackernagel and Rees subscribe to the view that the environment can no longer be viewed as a backdrop to human lives and action, but rather that humankind is embedded in greater Creation.

> The premise that *human society is a subsystem of the ecosphere*, that human beings are embedded in nature, is so simple that it is generally overlooked or dismissed as too obvious to be relevant. However, taking this "obvious" insight seriously leads to some profound conclusions. The policy implications of this ecological reality runs much deeper than pressing for improved pollution control and better environmental protection, both of which maintain the myth of separation. If humans are a part of nature's fabric, the "environment" is no mere scenic backdrop but becomes the play itself. The ecosphere is where we live, humanity is dependent on nature, not the reverse. Sustainability requires that our emphasis shift from "managing resources" to managing *ourselves*, that we learn to live as part of nature. Economics at last becomes human ecology.[53]

What makes part of Wackernagel and Rees' approach so powerful is that Ecological Footprint analysis is dominated by concrete economic analysis. While numbers, like words, can be manipulated, the indisputable facts that emerge from an Ecological Footprint analysis powerfully challenges accepted norms of behavior and consumption.

> Ecological Footprint analysis is an accounting tool that enables us to estimate the resource consumption and waste assimilation requirements of a defined human population or economy in terms of a corresponding productive land area. Typical questions we can ask with this tool include: how dependent is our study population on resource imports from "elsewhere" and on the waste assimilation capacity of the global commons?, and will nature's productivity be adequate to satisfy the rising material expectations of a growing human population . . . ?[54]

The Ecological Footprint goes far beyond merely calculating human rates of consumption and pollution outputs. The Ecological Footprint is concerned with all the different ways that land is being or not being used. For example, the Ecological Footprint serves as a means of calculating such critical elements as carbon sinks—that part of the ecosphere able to absorb the carbon dioxide (CO_2) released by fuel consumption.[55] The Ecological Footprint calculates both their distribution and disappearance.

Wankernagel and Rees remind their readers not only of the Ecological Footprints of our respective regions and nations but also what they refer to as "Fair Earthshares," those portions of the productive and inhabitable Earth which, if divided evenly among the planet's entire human population, each inhabitant would receive.

> Our ecological footprints keep growing while our per capita "earth shares" continue to shrink. Since the beginning of this century, the available ecologically productive land has decreased from over five hectares to less than 1.5 hectares per person in 1995. At the same time, the average North American's footprint has grown to over four hectares. These opposing trends are in fundamental conflict: the ecological demands of average citizens in rich countries exceed per capita supply by a factor of three. This means that the earth could not support even today's population, 5.8 billion, sustainably at North American material standards.[56]

Our own individual "earth shares" vary according to a variety of factors, such as hemisphere, nationality, class, ethnicity, race, faith, and gender. One clear problem among many is the relative "normalization" of these gross inequities.

> The earth is one but the earth is not. We all depend on one biosphere for sustaining our lives. Yet each community, each country, strives for survival and prosperity with little regard for its impacts on others. Some consume the earth's resources at a rate that would leave little for future generations. Others, many more in number, consume far too little and live with the prospects of hunger, squalor, disease, and early death.[57]

Doing Ecological Footprint Analysis

One key premise of Ecological Footprint analysis is that the terms it utilizes cannot be expressed monetarily. This is because monetary units naturally distort the true value of what Wackernagel and Rees refer to as "natural capital."[58]

> Natural Capital refers to any stock of natural assets that yields a flow of goods and services in the future. For example, a forest, a fish stock or an aquifer can provide a harvest or flow that is potentially sustainable year after year. The forest or fish stock is "natural capital" and the sustainable harvest is "natural income."[59]

At the same time, the monetary value of an item does "not distinguish between substitutable goods and complementary goods."[60]

> Moreover, on monetary balance sheets, all prices are added or subtracted as if goods that are priced the same are of equal importance to human life— money equivalency equates the essential with the trivial.[61]

Thus, according to Wankernagel and Rees, "prices do not monitor stock size or systems fragility, but only the commodity's short-term scarcity on the market."[62] This problem is compounded by the fact that "the potential for growth of money is theoretically unlimited, which obscures the possibility that there may be biophysical limits to economic growth."[63]

How does the Ecological Footprint analysis work in the absence of monetary measurements? It begins by measuring what Wackernagel and Rees refer to as "productive land," and compares that to the total number of hectares within the borders of a nation-state. The overall calculation includes both "productive" and "unproductive" land, which is variously assigned to eight categories. First, the Ecological Footprint analysis considers the percentage of the land that is no longer ecologically viable. This category is referred to as "energy land" or land appropriated for fossil energy use. The next category assesses "ecologically productive land," which includes gardens, cropland, pastureland, and managed forests. Next, Wankernagel and Rees refer to "consumed land" or land that has been built upon, sometimes referred to as "degraded land." Finally, there is "land of limited availability." This land includes untouched forests and nonproductive areas such as deserts and ice caps.[64]

In order to calculate the Ecological Footprint of a nation in terms of its land, one must begin by calculating the hectares of energy land (or land emitting CO_2) relative to the total hectares of carbon sinks and their ability to absorb CO_2 produced within the borders of the nation-state. One must then calculate the hectares of built-up land or "consumed land" relative to the total hectares of the country. These same calculation patterns are in turn applied to the other types of land already identified, in order to determine the percentage each category occupies relative to the total size of the nation-state.

How the land is used, not used, or degraded is only a portion of the picture, however. In order to calculate the Ecological Footprint of a country one must also consider the average consumption patterns of individual members of the population. These consumption patterns are divided into five categories: food, housing, transportation, consumer goods and services.

In terms of calculating the consumption patterns among the national population, one must first estimate the average person's annual consumption of the five categories (food, housing, transportation, consumer goods and services) and then divide the total consumption by population size. "For many categories, national statistics provide both production and trade figures from which trade-corrected consumption can be assessed."[65] The calculation looks like this:[66]

Trade corrected consumption = production + imports − exports

The next task is to estimate the "land area appropriated per capita (aa)" for the production of the five principal consumption categories (i).[67] This is done by dividing the "average annual consumption of that item as calculated above ('c' in kg/capita) by its average annual productivity or yield ('p' in kg/ha)."[68] The resulting calculation goes as follows:

$$aa_i = c_i/p_i$$

The next step is to compute the total Ecological Footprint ("ef") of the average inhabitant of the nation-state—"i.e. the *per capita* footprint"—by totaling all the "[ecosystems] appropriated (aa$_i$) by all purchased items (n) in his or her annual shopping basket of consumption goods and services."[69] The calculation appears as follows:

$$ef = \Sigma aa_i$$
$$i = 1 \text{ to } n$$

Finally, one "obtains the ecological footprint (Efp) of the study population by multiplying the average per capita footprint by population size (N)."[70] The equation is expressed in this manner:

$$Efp = N(ef)$$

"In some cases where the total area used is available from national statistics, [one] computes the per capita footprint by dividing the population."[71]

Clearly, there is ample room to be far more nuanced in calculating the Ecological Footprint of an individual or population. It has been suggested by Wankernagel and Rees that if one was to more accurately calculate one's own footprint, one would have to begin by weighing one's own garbage over a year-long period. On a broader scale, one must consider the fact that many consumer products require the use of a multiplicity of materials and energy sources. Thus, there is ample room to become far more nuanced than these initial equations suggest. What Wankernagel and Rees are suggesting is rather a standardized approach that allows "general case" comparisons among regions or countries.[72]

Realism and Ecological Realism

In light of our consideration of religion and international affairs, Ecological Location, and the Ecological Footprint analysis, classic Realism demands to be seen in a different light. The work of Burnett, Luttwak, Johnston, Spencer, Wackernagel and Rees challenges a variety of Realism's steadfast assumptions. Together, they define a new Realism, one that is eco-centric rather than anthropocentric and challenges the viability of classic Realism's understanding of the nature of power. In our new understanding, power itself has changed in character completely, and is no longer an entity to be projected in a self-serving manner. Rather, power is found in a nation-state's ability to protect, cultivate, and efficiently utilize domestically held natural capital, living out a type of sustainability that does not rely exclusively on trade in order to insure survival. Trade, in turn, becomes a matter of exchanging surplus natural capital with one's neighbors. Thus, while classical Realism's old paradigm of power may well have involved the cultivation of the ability of one or more nation-states to manipulate others, the new paradigm emphasizes the ability to control one's own consumption and pollution patterns and to impart one's knowledge of how to do so to any nation who has yet to reach a point of ecological equilibrium. Likewise, while Realism speaks of nation-states' responsibility to acquire, retain, and project power while acting out of their own specific self-interest, the borrowing or theft of others' sustainability is revealed as a very short-term accomplishment.

Under the construct of what I propose to call *Ecological Realism*, classical Realism's call to distinguish between the desirable and the possible is replaced with the distinction between the sustainable and the unsustainable. For this reason, a new definition of balance of power is in order—one that defines balance as ecological equilibrium, first within the borders of the nation-state, and then moving into ecological balance with one's regional neighbors. Thus while for classical Realism national survival is paramount, the principles of Ecological Realism (or Eco-Realism), claim that in fact it is global survival that is the real goal. For this reason, Eco-Realism holds that true self-interest must always be grounded in mutual interest. In this respect, Eco-Realism suggests a level of mutuality and permanent cooperation among nation-states, the likes of which traditional Realism could not have conceived. We must reject classic Realism's view that the balance of power must be seen as temporary and even unnatural. A new Realism is necessary, because diminishing resources and shrinking earth shares demand that nation-states cooperate on a permanent basis at a level of intimacy not previously conceived, or they will perish.

Classic Realism holds that the capacity to know what is good and evil in the realm of international relations is impossible.[73] Eco-Realism directly counters such a notion by arguing that what is good promotes the sustainability of the human and nonhuman members of the biosphere, and what is evil is whatever undermines the capacity for building, sharing, and maintaining sustainability. In this regard, the moral dimension of international relations can no longer be dictated by interest defined as power. Rather, Eco-Realism invites its practitioners to see that sustainability which is built mutually across nation-state and, ultimately, hemispheric borders is an unmitigated good. Power therefore comes to focus not on high-priced commodities trading but rather on low-priced or even free tech transfer between the North and the South.

While a classic Realist would label the above notions as highly unrealistic proposals that go against nearly every law that governs the conduct of the nation-state, Eco-Realism argues to the contrary. Twentieth-century forms of classic Realism emerged in an era that was only beginning to consider the phenomenon of resource scarcity and the potential ensuing conflicts it could engender. Unequal distribution of raw materials for survival was seen as a matter to be resolved in the realm of bilateral and multilateral nation-state trade and interstate armed conflict. The thinking of the time was highly regionalized, most often focusing on the East–West conflict. Purely regional thinking, however, is a luxury the modern analyst can no longer afford. Damage to the ozone, a dwindling fresh water supply, or the diminishment of carbon sinks that serve vast transnational areas demand a more explicit type of equitable global thinking. Eco-Realism is one attempt to describe what such global thinking might look like, and to guide a paradigm shift that in our notions is not only of power, but also of accountability, responsibility, sovereignty, solidarity, community, religion, land, commerce, and diplomacy.

Power and Accountability

From the Eco-Realist perspective, a nation-state's ability to strike an equilibrium in its use of natural capital and disposal of waste is the primary means by which a nation's power is measured. Dependency on the sustainability of other nations will be a sign of tacit weakness and vulnerability and the mere capacity to acquire raw materials from far-flung regions will no longer be seen as a right, privilege, or valued ability. Such a transformation requires a new understanding of accountability that has yet to be seen in the international commons. Accountability to one's own national population is in itself a foreign notion to many national leaders, beyond the boundaries of maintaining state power. Meaningful accountability to other national populations is even more rare.

Eco-Realism rejects the anthropocentric thinking of classical Realism that holds that nearly all power worthy of the name has been designed, controlled, and propagated by human beings. Eco-Realism holds that the power of the biosphere is in many ways greater than the power any group of humans can muster. No missile or commodities exchange is capable of replacing the ozone or generating accessible fresh water. Classic Realism is convinced that the human is the pinnacle of the power chain. We now know that this is not so. In the biosphere's own changing patterns, the capacity for human adaptability comprises a critical type of power that cannot be matched or ignored. The biosphere, with its capacity for regeneration and its limited supply of materials for human exploitation, demands a new level of human respect, accountability, and material simplicity. As Spencer, Wackernagel and Rees have noted, the human is embedded *within* the biosphere and, therefore, can never claim outside observer status. Until now, many have seen human dependence upon the nonhuman members of the biosphere as an explicit sign of weakness. The time has now come to view such a relationship both as an advantage and as a primary source for lessons about future human eco-centric conduct. Human accountability must now embody, as Larry Rasmussen has written, "a turning to Earth."[74]

Sovereignty

The ecological crisis itself, while not necessarily or at least initially diminishing nation-state sovereignty, will demand a reappraisal of the meaning of sovereignty. Borders could be superceded by the reality of bioregions. Bioregions could act in concert, not toward classic Realism's goal of containment, but toward the goal of building the sustainability of natural capital and equitably sharing the knowledge gained by those in more marginalized locations.

Individual, communal, or state accountability to an entire bioregion will prove difficult to establish and maintain. Yet while current political borders will no doubt remain long into the future, they could potentially come to be seen more as historical distinctions rather than ecological realities. Already multinational corporations (MNCs) have shown that breaking down national borders is more than a pipe dream. In many cases MNCs have acquired a level of power that has directly challenged the ability of many nation-states to control their own domestic and international affairs. This "breaking down" of national borders could be embraced by proponents of Eco-Realism as an opportunity rather than a disadvantage. Classic Realism did not envision how porous national borders might become. The goals of Eco-Realism require such a porousness of the borders that divide nation-states, at the very least in terms of the perceptions of the national populations.

Community

Building communities across national borders is a priority for proponents of Eco-Realism. At the very least, Eco-Realism invites ethnic, religious, and otherwise culturally tied populations to see once again that they have often been divided by abstract borders for the benefit of others, just as regional ecosystems have been parted out along what are often illogical and destructive lines. In practice, the land itself and the methods by which people make their living from it often differ little on either side of national frontiers. This is because agricultural communities that are separated by national borders often mimic the neighbors they have in the same bioregion in their cultivation methods and products. In this respect, emerging transnational agricultural communities already exist on either side of national borders. Yet how can such often-disparate communities become more closely bound together? Therein lies a central question of Eco-Realism.

Clearly Eco-Realism acknowledges that the inherent power of national populations to shape their own futures is a matter their own national leaders often discount. If sustainability or equilibrium of natural capital is ever to be realized, then the full participation of people on the ground will have to be secured in order to reach this goal. In the North, this will entail the personal decision of millions to end their dependence upon the sustainability of others. The degree to which Northern leaders can coerce their respective populations to embrace sustainability is somewhat suspect. The current unwillingness of the United States to ratify the Kyoto Protocol is only the most recent example of a Northern nation placing its perceived "right" to pollute over the needs of the global community. Despite these facts, many ordinary citizens still have it within their capacity to alter their own consumption and pollution patterns. At the same time, there are many communities, particularly in the South, which are acting far more sustainably than many of their Northern counterparts have ever conceived. For this reason, it is more often the South rather than the North that should be looked to for lessons in sustainability.

Exchanging information on how to become sustainable will require populations to communicate with each other at a more sophisticated and intimate level than their national governments and circumstances often allow. People must be given opportunities to tell their stories about their ties to the land, both physical and spiritual, as well as their own understandings of responsibility to the Earth based on their cultural teachings and their faith. In many respects, such conversations will have to cross religious and cultural lines that have rarely been breached. Hearing and understanding these stories and experiences will be critical if national and regional populations are to have the chance to promote solidarity across existing borders. This may well be the first time national governments have made an effort to understand the impact of popular religion on their own populations.

Religion and Land

As Burnett, Luttwak, and Johnston have acknowledged, our disregard of religion in the realm of international relations has often blinded us to the facts on the ground. The fear of being somehow "unscientific" has left many analysts unable to understand the national populations about which they claim an expertise. Eco-Realism acknowledges that focusing on the religious beliefs, practices, and histories of the populations it seeks to make more ecologically sustainable is central to any effort at individual, national, and regional transformation. Such an approach embraces Spencer's placement of great importance on a population's Creation stories and ecological histories. What do the prevailing religious traditions of a people have to say about their perception of Earth's origins? Where did humans come from? What is the Divinely acknowledged human responsibility to greater Creation? Who are our neighbors and what are our true responsibilities toward them? What are the ritual practices that regionally bind groups of people together? What are considered by a particular people to be the most important feasts and festivals? Who are the sages and saints whose stories continue to draw people's attention? The seeds of what binds a community together can often be found within the answers to these questions. Conversely, these same seeds are as often the basis of conflict. For this reason Eco-Realism places a premium on the task of all parties who seek sustainability to familiarize themselves with the lived-out religious practices and beliefs of the people they seek to work with. In turn, fighting for the right of transnational populations to communicate along religious and cultural lines will also be a central task of the proponents of Eco-Realism.

Eco-Realism acknowledges the unlikelihood that such exchanges will take place easily. Too many factors stand against the possibility of such communications. Issues of national control, market forces that favor the continuation of existing paradigms, and perhaps the most important of all, the ingrained perceptions of the national populations themselves serve to prevent such exchanges. Employing creativity in effectively asking members of national populations to try to disregard long-held hatreds, be they religious, racial, gender-based, class-based, or ethnic, will be an extraordinary challenge in and of itself. It has often been to the advantage of national governments to perpetuate such bases of conflict in the name of retaining their own individual interpretations of "national security." The choices of national governments are thus a key factor in the transformation envisioned by Eco-Realism.

How can members of national populations telling their foundational stories (including their Creation stories) across existing borders change long-held human habits of consumption and pollution? On one level, the realization of the commonality of many people's lives across national borders will be a revelation for some. For others, the surprise will come in the form of the often complementary nature of their respective Creation stories, religious practices and beliefs. These phenomena could be the basis for further dialogue across transnational borders. Most importantly, the goals of Eco-Realism will require a spiritual transformation in the lives of a significant portion of national populations and their key leaders. In this case "spiritual transformation" means a turning to the Earth as a whole, not an Earth that is simply a provider of resources, or the host to multiple lines of demarcation. Such a spiritual transformation makes room for the belief that humans can learn to live in something other than an anthropocentric world—a world that is ecologically interdependent and deeply integrated where humans are but one species among many. Regretfully, many members of the human community have yet to realize that they are not always the most powerful members of Creation. The popular religious beliefs (especially the Creation stories) of the world's many religions are often better arbiters of such visions than any human.[75] This places the telling and sharing of such beliefs and practices across multiple borders in a place of great importance.

Religious NGOs also play an important role in helping to foment a higher level of eco-centric education and cooperation among transnational populations. The Geneva-based World Council of Churches (WCC) is just one example of an international religious NGO, which is involved in this work. The WCC's efforts surrounding the Kyoto Protocol are one example among many where a religious organization has played an important role in raising transnational awareness while confronting the secular arbiters of power regarding climate change.

Regretfully, many of those who claim to represent religious interests on an international level have proven to be self-serving advocates of positions that are often calculated to divide rather than unite people across divisions of race, geography, class, gender, and ethnicity. The spirit of Eco-Realism calls for the opposite state of affairs. By describing an undividable and deeply interdependent biosphere in which all humans are embedded, Eco-Realism invites religious actors to speak of an interdependent ecosphere as the Divine model for cooperation among human and nonhuman members.

Unfortunately for many of the Earth's human inhabitants, discussions of the GAIA hypothesis will lead to little practical reform. GAIA's suppositions (which are based on the premise that the Earth is one contiguous and interdependent biological unit) are often presented in too abstract a manner to capture the imagination. For this reason, focusing on land itself can be a way to draw a wide cross-section of people into a conversation. Often, the smallest villages in the world have religious festivals that are tied to the local land. In turn, there is often a vocabulary, be it through ritual, song, storytelling, or dance that reflects the ties that individual communities have to the land they inhabit. It is significant that popular religious ritual is more often an outdoor activity than something that takes place within the walls of a synagogue, church, mosque, or temple. In these rituals the Earth is often remembered, honored, celebrated, or understood through the eyes of the local inhabitants.

Eco-Realism acknowledges that the ties which people feel to the land where they live or to the land of their forbearers are often profound. The land is the provider of stories—stories of family, stories of hardship and plenty, myths, and stories of the Divine. While the land certainly is the holder of the ecological history that Spencer speaks of, it is also the gatekeeper he names of the spiritual histories of the people who inhabit it. People use God-language to talk about the land. Many people speak of God's blessings or lack of blessings when describing a harvest. Others speak of a fate that God allotted them that accounts for their easy or difficult experiences in trying to make a living from the land. Most, however, speak of a God who created the land, although many ascribe such notions to stories they were told in school rather than as an actual reflection of reality. Nonetheless, God-language persists in even the most secular environs, and it is often difficult for an outsider to determine to what degree people are being sincere or merely habitual in their use of such words and phrases. Eco-Realism, however, is not in the business of examining the degree of faith an individual or community of people possesses. Rather, Eco-Realism seeks to develop a hermeneutic of land as a means for discussing the Divine and for hopeful motivation to change consumption and pollution patterns. This is the work of Eco-Realism, which seeks to inspire the human community toward a different way of understanding itself and its relationship to the Earth. Such transformative change is difficult to imagine in the absence of speaking about the land and the different Creation stories that explain its existence and purpose.

Commerce

As it has been observed by Wackernagel and Rees, our current economic practices grossly distort the value of many of the commodities we trade.[76] Money, which is theoretically unlimited in supply, masks the reality of the increasingly limited quantity of natural capital that can be found on our planet. At the same time, the phenomenon of globalization has only increased (particularly the North's) capacity to borrow or steal the sustainability of others. Eco-Realism advocates the notion of subsidiarity, which holds that goods and services should always be drawn first from the most local locations possible for the consumption by the local population, and that goods and services which come from great distances should be sought out only in exceptional cases. Enacting such profound changes will be extraordinarily difficult. Currently, we face an increasingly globalized market economy that often concludes that it is cheaper to transport fruits and vegetables from thousands of miles away rather than grow them locally. So, too, has the disparate cost of human labor drawn businesses to sever ties to local laborers in favor of low-wage, low-rights workers in other countries. While the transportation costs of current trade practices are acceptable to many Northern actors, future levels of petroleum reserves may require at least a partial return to the dictums of subidiarity and limits on the use of fossil fuels.

While it is easy to critique current practices of international trade, it is much more difficult to persuade the market's dominant forces to radically alter current practices. While the dictates of Eco-Realism call attention to many city's, region's, and nation's true earth shares and the long-term consequences they engender, such admonitions will most often fall on deaf ears. Market actors who once prided themselves for their long-term thinking have given way to a new generation who is often only interested in the latest quarterly earnings report. Long-term thinking is less quickly rewarded, and so it is far less practiced. Eco-Realism demands a long-term view, despite the fact that fewer in the marketplace care to grasp its implications.

Do MNCs have contingency plans in the event that the polar caps melt? Is it true that there are already those who are trading in futures that involve clean air and fresh water? It is a strong possibility that however sobering projections for the future of the ecosystem become, many of the actors who guide the market intend to take everything in stride, while altering little if any of their approach to trade. If this is true, then the proponents of Eco-Realism must focus their efforts not exclusively on those in positions of national and international power, but on a more local level.

Eco-Realism seeks to support local communities that manufacture local goods with local labor and local materials for local consumption whenever possible. Such sustainable practices are more often found in the South than in the North. This goes beyond the fact that most Southerners simply consume less than their Northern counterparts and move into the realm of those Southerners who actively fight against the globalization of their local communities. For this reason, it is likely that Southerners will teach Northerners how to live within their Fair Earthshares while lessening their own Ecological Footprints as they seek sufficiency.[77] Yet it will be difficult for many Northerners to accept Southern assistance. For many Northerners, former Southern colonies are better suited to act as the point of extraction of key raw materials and a source of cheap labor. To honor the South as a teacher will prove difficult for many. Nonetheless, this is exactly the sort of perceptual shift that will be required in order to realize the goals of Eco-Realism.

Eco-Realism and Sustainable Diplomacy

How can the goals of Eco-Realism best be realized? To what degree can one expect to work within the existing system? At its base, Eco-Realism requires the creation, promotion, and maintenance of a new set of domestic and international relationships. As Wackernagel and Rees point out, our current perception of the nation-state as one of the primary arbiters of influence must ultimately be rejected in favor of the reality of the bioregion. For this reason, the type of diplomacy required by Eco-Realism will differ from what is currently practiced. The inefficiency of singular state-to-state relations will have to give way to focusing on bioregional ties. This is because while singular state-to-state relations cannot be ignored, they cannot stand as the main bulwark of an eco-centric diplomacy—or what I will call *Sustainable Diplomacy*.

Religion, Land, and Power

Sustainable Diplomacy must be initially understood as the praxis of Eco-Realism in state-to-state and region-to-region relations. Its goal is to ultimately de-emphasize state-to-state relations in favor of promoting and maintaining relations between bioregions. For this reason, Sustainable Diplomacy is a process of fomenting relations, which go well beyond exclusive contacts between human elites. Sustainable Diplomacy therefore places a high value on understanding the lives of the populations it is affecting. In order to develop such understanding, Sustainable Diplomacy begins by mapping the terrain that links religion, land, and power in both human and nonhuman populations. For this reason, the practice of

Sustainable Diplomacy requires a more profound understanding of the lives of those living "on the ground" than many schools of diplomacy advocate.

The current configurations of diplomacy as they are practiced will not disappear tomorrow. Thus, to understand how Sustainable Diplomacy will work on the ground, it must be placed as a transparency over our current practices. Without engaging in this exercise, it will be difficult, if not impossible, to envision how we can accomplish a transition from the current state of affairs to the practice of Sustainable Diplomacy. We must ask the following questions: (1) what new questions will arise in the course of this transition?; (2) how must our current practices and underlying suppositions change in order to move toward Sustainable Diplomacy?; and (3) under Sustainable Diplomacy, what will remain of our current methods of practicing foreign policy? These are a few of the guiding questions that the practitioners of Sustainable Diplomacy must ask while engaging humans, nonhumans, and bioregions on the ground.

The work of Johnston, Luttwak, Berry, and Burnett guide the practitioners of Sustainable Diplomacy to recognize that religion can often play a central role in the promotion of conflict and conflict resolution. For this reason, diplomats who choose to practice Sustainable Diplomacy will have to be more than passingly familiar with the faith traditions as they are practiced by human populations that inhabit the nation-states and bioregions with which they are interacting. Sustainable Diplomacy requires that one understand how states can use religion to retain domestic and sometimes regional dominance. In turn, the advocates of Sustainable Diplomacy will have to acknowledge how a difference in faith traditions between two countries or bioregions alter their mutual perceptions of one another, and can affect their willingness to work together toward a common goal. This can be true at the elite level, as well as in relations between respective populations. For this reason, Sustainable Diplomacy's acknowledgment of the importance of religion in international relations not only interprets the present, but can also be predictive of the future.

Just as Spencer's Ecological Location has advocated, practitioners of Sustainable Diplomacy must seek to understand the links between the religious language and images used by the people of a region to convey place and meaning. For many, ties to the land cannot be described without also describing ties to the Divine. These are often inseparable. At the same time, the beliefs central to a people cannot be completely separated from the elites who emerge from among their population. Knowing this, sending a representative who lacks knowledge of the regional faith traditions must be seen as the equivalent of sending an individual who is unable to speak the language of those with whom he or she hopes to communicate.

The land, as Spencer has noted, shapes the perceptions, languages, and beliefs of those who inhabit it. Therefore, any practitioner of Sustainable Diplomacy must place a great emphasis on learning how the land feeds or does not feed a regional or national population, both literally (in terms of food) and spiritually. Such a responsibility will require an intimate knowledge of regional farming techniques, waste disposal methods, and national and regional efforts that are working toward the ability to produce and maintain an adequate amount of sustainable natural capital to support the human and nonhuman population. At the same time, the practitioner of Sustainable Diplomacy must strive to learn about the Creation stories and other popular religious beliefs that are taught among the people and drawn from their faith traditions. This is because Sustainable Diplomacy holds that it is within the spiritual richness of these stories and traditions where one can often find the seeds for advocating a new level of sustainability and accountability among people.

As in the case of Eco-Realism, Sustainable Diplomacy recognizes that long-term power rests in the ability of a population to strike an equilibrium between its use of natural capital and its disposal of waste. This is not to say that current methods of projecting power are going to disappear overnight. The use or threat of armed force, monetary pressure, and other strategic uses of intimidation will no doubt remain with us far into the future. The task of Sustainable Diplomacy is thus not to dwell exclusively on these traditional projections of power, but rather to help cultivate alternative extant sources of power and influence in the bioregions that might accompany and eventually overtake existing practices. In order to move toward these goals, the practitioners of Sustainable Diplomacy must become expert networkers, helping to bind formerly opposing populations together under the banner of long-term sustainability and ecological survival. Such aims can only be achieved if more people come to realize the truth of Wackernagel and Rees' conclusions: monetary value does not necessarily equal survival value, and no amount of strategic arms can permanently postpone a sobering reckoning regarding the survivability of the ecosphere. For this reason, those who advocate Sustainable Diplomacy must become a far-reaching conduit for those who tell the Creation stories, for those who celebrate popular feasts and festivals, for those who advocate for the sacredness of Creation, and for those whose faith has compelled them to turn to the Earth.[78] Power, therefore, also comes in the form of sharing popular religious beliefs and practices across existing frontiers, and privileging voices that have until now been marginalized. Of equal importance is that all these tasks must be performed across boundaries that are too infrequently breached. Thus the Muslim and the Christian must share instead of imposing their interpretation of Creation stories on their Jewish neighbors. Northerners must sincerely listen to the wisdom of Southerners. And those who have been divided by race, gender, or ethnicity must be invited to the same table, to share in the same exchanges Sustainable Diplomacy envisions for any other group. While experience would show that such exchanges are less than likely to occur in our current circumstances, it is the work of Sustainable Diplomacy to visualize, persuade, explain, and attempt to enact that which others refuse to give voice to.

Building and Maintaining Sustainable Communities

Building and maintaining transnational sustainable communities is one of the ultimate goals of Sustainable Diplomacy. In order to begin this work, proponents of Sustainable Diplomacy will have to identify the principal existing points of conflict between the nation-states or bioregions they are attempting to bring together. In addition, it will be critical to identify the existing communities that are already sustainable. Finally, those who promote the agenda of Sustainable Diplomacy must ask the following questions: (1) how might new means of building sustainable communities lessen or eliminate existing points of conflict?; (2) what are the existing sustainable practices that are currently being employed by the respective populations?; and (3) who are the individuals or groups best suited to share information across borders regarding these existing sustainable practices?

The task of maintaining existing sustainable communities while building new ones is an ongoing work. The mobility of populations, the lessening of individual earth shares, long-term habits, and the current market forces of globalization all work against the realization of true sustainability. For this reason, Sustainable Diplomacy's advocates (particularly in the North) must be willing to open their eyes and see that there are in fact many communities, most often in the South, who are already living relatively sustainable lives. The model of these communities must be lifted up as realistic examples of sustainable communities for those who lack the ability to see that there is another way of living.

Sustainable communities are built from the ground up, first within individual communities and then between communities.[79] The architects of Sustainable Diplomacy must therefore be willing to engage individuals and small groups on the local level, while maintaining a dialogue with the elites. At the same time, Sustainable Diplomacy's advocates must keep in mind that social and environmental justice are integral to one another.[80] In this case it is critical to remember that "justice" is not a synonym for simple numerical equality; it is a collective *mutuality* in which "we share one another's fate" and "promote one another's well being."[81] The starting point of sustainable community is therefore found in the act of entering into the predicaments of those who suffer, for compassion (suffering with) is the passion of life itself.[82] Justice is served when our communities live out this tangible compassion while they embrace the normative guideposts or markers of Sustainable Diplomacy that have been illustrated in this chapter: *solidarity, participation, sufficiency, equity, accountability, material simplicity, spiritual richness, responsibility*, and *subsidiarity*.[83] Sustainable Diplomacy is not possible without these elemental foundations.

Larry Rasmussen notes, "sustainability owes as much to its socio-ethical character as it does to its technical prowess and knowledge base."[84] Sustainable communities will therefore not survive through exclusively material means. They will only endure when the "social and cultural health of communities is given the same level of attention now lavished on their technological and economic development."[85] This is because "our current lack of sustainability is in many respects a crisis of culture—a culture that is unsustainable, in part due to our collective unwillingness to submit human power to grace and humility."[86]

Advocates of Sustainable Diplomacy must strive to make clear that the human economy is only a subset of Earth's economy. This is clearly a paradigm shift of the first order. Yet without this shift, the human community will continue to shrink its earth share, lose sight of those communities who are practicing sustainability, and fail to grasp that the crisis it faces is equally spiritual as it is material. It is for this reason that the practice of Sustainable Diplomacy will involve a different manner of representing one's own government and bioregion. Practitioners of Sustainable Diplomacy will not only share in the political, economic, and consular duties current diplomats undertake, but they will also be conveyers and receivers of culture—including the stories of marginalized peoples and lands. They will become co-teachers of sustainability. At the same time the Sustainable Diplomat will become a more profound type of listener—seeking connections on the ground and across borders, between peoples, faiths, agricultural practices, and pollution management.

CHAPTER 2

THE FOUNDATIONS OF THE ECO-HISTORICAL LANDSCAPE OF MOROCCAN–SPANISH RELATIONS

In many respects, history is simply collection of subjective impressions. Our own historical perceptions can be influenced by our geographic locale, our class, gender, race, age, and education. In the name of agreement, we often focus on dates, movements, and personalities in order to arrive at a version of history that we are comfortable calling "relatively objective." Nonetheless, history as a rule is most often a compendium of facts written by the conqueror, rather than the conquered. For this reason, history is often Euro-centric, economically driven, and politically calculated to place those who have "won" conflicts in the best possible light.

As envisioned by Daniel Spencer, an Ecological History provides a decisive break from past historical methodologies, though it is hardly free of subjective interpretations. However, rather than concentrate exclusively on the exploits of one particular species (i.e. the human), an Ecological History embraces new and precedent-challenging elements, while embedding human history within its folds. This is because Ecological History is tied to the land, to the greater biosphere, and to all those who live within it. Ecological History is thus a story of abuse and respect for the land by its inhabitants. It is also a tracking of critical forks in the road that have led to sustainability and unsustainability. At the same time, an Ecological History is one that honors the Creation stories and subsequent theologies of those who live within the region being considered, be they competing or complementing. For this reason an Ecological History must by its very nature engage faiths and cosmologies, to ultimately present the theological anthropologies of the peoples who inhabit the land.[1] Ecological History tries to map what happened to the land and its nonhuman inhabitants alongside any and all human exploits. Thus, the actors of Ecological History are far more numerous than those of its conventional counterparts, and its stage is far wider and more inclusive. Such is the foundation of the worldview of Ecological Realism and Sustainable Diplomacy; it charts the ecological reality of the human and nonhuman lives on the ground, while putting the work of statesmen and stateswomen in a more realistic Earth-centric perspective.

Let us then begin our examination of the ecological and religious histories of Spain and Morocco with excerpts from a series of interviews that were designed to invite a variety of Spaniards and Moroccans into a discussion of their own Ecological Locations. The first interviewee, Javier Martos, is a young Spanish teacher who works in the Andalucian capital of Seville, his family's roots are in central and northern Spain. The second subject, Sadik Rddad, is a professor of English Literature at the University of Fes, whose birthplace is far south near the Sahara. These interviews, like others that will follow, are an attempt to place the voices of Moroccans and Spaniards directly into dialogue with every text and topic this book will engage. The voices of all the interviewees are therefore central in establishing the foundations necessary for building and maintaining Sustainable Diplomacy.

Javier Martos Cando

What do you think the land here was like when only the Iberians lived here?

I think that it was more fertile since humans were more directly connected to the land. In fact, in old Spanish culture, the primitive ones like the Iberians, they had a special connection to the land, something similar to the farmers nowadays. They knew that they had to take care of the land in order for the land to respond, then they respected it more than now because they depended on it. Moreover, land was some kind of divinity to them. I think that, in some way, people in the country (now) think the same, they see the land as a living asset. Then, I'd say that the land was less populated, it was a younger land, and there was a deeper relation between the two. People knew that the land was something necessary for living and, therefore, they took care of the land, and land that is well cared for is fruitful. I think that this is very important, moreover, in that at the time there were no big cities, there were primitive settlements, small groups where the connection between land and population was a double one. Land for them was like their mother, the symbol of fertility and prosperity. If the land failed, they died.

What happened to the land when people came here from other places?

I think that two things happened. On one hand, it was something positive because . . . .when some other people came . . . as I told you before, Spain was and still is, a very fertile land . . . .then some others came . . . .it was the beginning of the land exhaustion. Why? Because when other cultures came here, for example the Romans and all the ones after them, [they] tried to exploit the land to extremes. They squeezed the land. So, it was positive in the sense that the land produced what it had to produce because it was correctly exploited. But this is also a negative thing because, as I told you before, the Iberians respected the land a lot because they knew that they [depended] on it, but when others came, they began looking at the land as a wealth source to be exploited and [these outsiders] lost their respect for the land.

Javier Martos Cando, 33, teacher, Seville[2]

Sadik Rddad

What do you think the land was like when only the
Imazighn lived here?

*I would think it was . . . Imazighn as I am I wouldn't try to look at it . . .; in my
imagination I look at it as a utopia. But I would imagine that there were people
killing each other, there was a lot of tribalism. And the tribes would kill each other
for water, and for domination. But the major difference, of course there are major
differences, the major difference is that the Iamazighn are people who live very,
very close to the land. People who live in the mountains. I don't know of any monuments left by the Imazighn, so I would think that the Imazighn lived in tents . . .
they were [nomadic] rather than sedentary tribes who didn't care too much about
living in cities, but they lived very, very close to the land, very close to the water.*

What do your religious or cultural traditions teach you
about the origin of the land?

*Religiously speaking . . . before, nothing was there. In six days, God created this
world because he wanted to create someone in his own image to shape him. And so
he created the land. Now, I'm not looking at it from a scientific point of view, but
from a religious point of view. There was nothing, so God wanted to create something. I don't know why, but . . . maybe to ascertain his existence, whatever. God
wanted the land to be, and it was. It's a creation of God. I believe it—that it was a
creation of God. And that's it.*

Sadik Rddad, 39, professor, Fes[3]

Spain and Morocco: An Eco-Historical Portrait

Level One: Geology, Geography, Genetics, and Linguistics

To consider Spain and Morocco from a geological and geographic stand-
point is to note numerous similarities. Separated by less than 13 km by the
Strait of Gibraltar, a great deal of the Iberian peninsula was formed at the
same time as its southern African neighbor, Morocco's northern coast.
The lands (which nearly touch but for the Strait) were formed during the
Mesozoic, Tertiary, and Quaternary periods. In fact, these two continents
originally touched one another, perhaps as recently as 65 million years
B.C.E, prior to continental drift. The lands of southernmost Spain and
northernmost Morocco came into existence between 225 and 2 million
years B.C.E and are relatively young in comparison to interior lands on both
continents.[4] The areas that comprise modern Morocco and Spain once
held equally significant deposits of iron, lead, and zinc.[5] Spain, however,
also originally enjoyed significant deposits of gold, silver, and copper, while
Morocco boasts the world's second largest deposits of phosphates, an
attribute that did not seriously impact the economic wealth of the country until the modern age. Clearly, the original human occupants of the

Iberian Peninsula enjoyed mineral advantages not known by their closest North African neighbors.[6]

Spain

Modern Spain occupies 80 percent of the Iberian Peninsula situated in southwestern Europe, covering 50,452,006 ha.[7] The Iberian peninsula is bordered by France to the northeast but is blocked from easy passage by the formidable Pyrenées mountain range, which offers four possible routes of travel among peaks that exceed 3,048 m. Spain shares the Iberian Peninsula with Portugal to the west, retains control over the Canary Islands (in the Atlantic off the coast of southern Morocco), the Balearic Islands in the Mediterranean, and two significant enclaves on the Moroccan landmass: the cities of Ceuta and Melilla. Spain is less temperate than most areas of western Europe, having hot summers in most regions, with temperatures averaging about 35°C. The mountainous areas in the interior experience cold winters.[8] The Spanish landscape hosts a variety of ecosystems, including the lush rain-soaked mountainous region of Galicia, the high plains of the interior (the second highest point in western Europe), a rich, vast, and varied collection of coastal ecosystems, and the often rocky lands of the southwestern region of Extremadura. Spain boasts a significant portion of arable land, which currently comprises 31 percent of the nation's landmass.

It is certain that Spain's original human inhabitants lived in one of the richest ecosystems in all of western Europe, which is evidenced by the fact that even today postindustrialization Spain hosts some 5,000 species of plants, Europe's most diverse variety of flora.[9] Spain's original forests were also significant in size and variety, including large areas covered by hardwood forests of oaks (Cork, Pyrenian, and Holm), pines (Stone and Aleppo), and numerous Beech, Scotch pines, Tamarisk, Willow, Black Poplar, Ash, and Alder.[10] The presence of Ash and Alder (which have since disappeared from the peninsula) indicates that the early Iberians lived in a more humid climate than modern Spaniards.[11]

Spain's original fauna were also quite extraordinary and included Timber Wolves, and the Iberian Lynx. Iberian birds included the Hermit Ibis, Adalbert's Eagle, the White Tailed Eagle, the Barbary Falcon, the Gyr Falcon, the Peregrine Falcon, the Bouibara Bustard, and the Slender-Billed Curlew. The waters off the coast also teemed with fish and ocean-going mammals, including Eurasian Otters, the Common Sturgeon, Mediterranean Monk Seals, Blue Whales, Minke Whales, Northern Right Whales, Sardines, Fin Whales, Sei Whales and Sperm Whales, Squid, Octopus, Flounder, Cod, a variety of Mollusks and Crustaceans, as well as a variety of Eels. Not to be forgotten are the Wild Boar and a precursor to modern cattle.[12]

The original human inhabitants of the Iberian Peninsula, known as the *Iberians*, are a topic of much academic dispute. While some called them "savages," others consider the Iberians to have produced one of the true "high cultures" of western Europe, as evidenced by their art and other artifacts.[13] The first evidence of humans on the Iberian peninsula comes from the late Paleolithic or Old Stone Age, beginning about 20,000 B.C.E., when stone tools were first used by humanoid creatures, and ending around 7,500 B.C.E.[14] The oldest evidence of humanoids occurs in the northeastern section of the Peninsula among those who came to be called the *Basques*. In southern Spain, the evidence suggests that the first extant human settlements arose in the first half of the early eighth century B.C.E in the southern region of what we now call Andalucia, which was originally called Tartessos. These people lived in structures made of walls plastered with lime and floors paved with red clay.[15] Concurrent with the final Bronze Age, among these dwellings there is evidence of wheel-made pottery and kilns of high temperature, signaling new and important social changes among the Iberians.[16]

While our current level of information regarding the religious practices of the Iberians is sketchy at best, archeological digs have rendered a fascinating combination of facts and speculation. According to J. Donald Hughes, the early Iberians, much like their Gaulic neighbors to the north, were originally hunter-gatherers who saw nature as the source of life.[17] Hughes hypothesizes that the early Iberians developed rituals that attempted to assure the success of the hunt. Evidence for these rituals is found in the cave paintings of Altamira and Puente Viesgo.[18] Depicted in some of these paintings are rituals in which, notes J. Donald Hughes, "both magic and religion were present and not yet differentiated."[19] According to Hughes:

> All their ritual practices seem to have been conducted with nature symbolism. Sacred dances in which men wore the heads and horns of animals are depicted on cavern walls in Mediterranean France and Spain, and the evidence indicates they used animal motifs in the ritual for initiation of youths into manhood. Modern primitive [*sic*] hunters are known to pray to and to propitiate animal and plant spirits and to adopt their representations as identifying and protective totems. They seem also to have a general feeling that they hunt and gather from need only, and should not kill or destroy wantonly. They sense that the other creatures of the earth have spirits and consciousness, and the power to help or hurt. This attitude should not be lightly dismissed as "anthropomorphism" or "the pathetic fallacy." It is a belief which has a genuine function in the delicate relationship between these people and the ecosystem within which they live. At their level of technology, they are indeed subject to the power of nature, and if they fail to find the proper balance within it, they will suffer.[20]

Juan Lalaguna reports that it is clear from some of the more recent archaeological evidence that early Iberian religious practices were nature-based, focusing on the importance of animals in ritual.[21] Nearly life-size stone statues that have been unearthed in Alicante, Murcia, and Albacete depict people and animals, along with small teracotta and bronze votive offerings.[22] According to Lalaguna these artifacts "show a strong eastern Mediterranean influence and perhaps [evidence the presence] of animal-based cults."[23]

By the middle of the fifth until the third century B.C.E we have evidence of cereal cultivation among the Iberians, consisting of spelt, hard wheat, and barley.[24] There is also evidence of a variety of domesticated legumes: "peas, lentils, chickpeas, broad beans and vetch."[25] In addition there is evidence of olives, though it is difficult to discern whether they were wild or cultivated, along with grape vines.[26] Ultimately, the Iberians became cultivators of livestock, which served as a source of milk and meat and as beasts of burden, as is evidenced by the preponderance of chicory (a popular fodder for cattle) at various digs.[27]

Morocco

Located on the northwest coast of Africa, Morocco has a long western coastline on the Atlantic Ocean and a northern coastline on the Mediterranean Sea, and covers 710,850 sq. km.[28] Morocco is bordered on the east by Algeria and on the south by either the western Sahara, or Mauritania, depending on one's political outlook.[29] Morocco's northernmost coast, near Tangiers, ends at the Strait of Gibraltar, which on a clear day affords a view of the Spanish coastline, less than 13 km away. Morocco's geography is composed of highly varied ecosystems, boasting four major mountain ranges, many of which are covered in forests and snow, a rather arid interior, a long, rich, and varied coastline, and a significant portion of the Saharan desert. The northern portion of the country holds a large amount of arable land, though it comprises only 19 percent of the country at present. The Moroccan climate is considered to be semitropical, with intense heat to be found in the summer in the interior plains, while the coasts are often cooler but sunny year-round.[30]

The Moroccan ecosystem was remarkably rich when considered in its original (prior to the arrival of humans) state. Even in the modern era, geographers divide Morocco into 23 distinct ecological zones.[31] It included significant forests made up of a variety of Pines, Cork Oak, as well as trees unique to the country such as Thuya and Argan. Even today, Morocco contains all bioclimactic zones found in the Mediterranean region.[32] There are a variety of rivers that flow from the mountains during the winter months, providing fertile areas for cultivation. Climate variations provide distinctly different ecosystems, as the northern coast is often warm and dry, while its winters are cool and wet. At the other end of the country, the beginnings of the Sahara, the world's largest desert, are given to extreme temperatures of hot days and freezing nights. Overall, one can divide Morocco into three major ecological regions: "the Rif and the Atlas mountain chains, the coastal and interior plains . . . and the semi-arid pre-Sahara of the

South."[33] According to Dale Eikleman,

> The first two of these regions show little marked discontinuity with the southern half of Spain, a resemblance complemented by others in traditional rural *genres de vie* and in the spatial organization of traditional towns.[34]

Such physical similarities between Spain and Morocco do not end here, as we shall soon see.

Morocco's original fauna were also rich in variety and included the Barbary Lion, the Northwest African Cheetah, the Caracal, the Leopard, the Eurasian Otter, and the Addax. The land also supported the Giant Buffalo, the Elephant, the Barbary Sheep, the Rhinoceros, and the Hippopotamus.[35] In addition, there was the Ostrich, the Dalmatian Pelican, the Hermit Ibis, the Imperial Eagle, the Barbary Falcon, the Peregrine Falcon, the Houbara Bustard, and the Slender-billed Curlew.[36] The reptiles were also well represented, and included the Baghdad Small-grain Lizard, the Green Turtle, the Hawksbill Turtle, and the Loggerhead.[37] Like Spain, the fish and mammals that swam off Morocco's coasts were rich and plentiful, including the Mediterranean Monk Seal, the Northern Right Whale as well as Sardines, Squid, Octopi, Flounder, Cod, and a variety of Mollusks, Crustaceans, and Eels.[38]

The first evidence of humans in northern Morocco (or what was originally referred to as the northwestern portion of the Maghreb, an area that included modern day Algeria, Tunisia, and Libya) appear from 20,000 to 7,500 B.C.E, and are referred to as *Iberomarusians*.[39] This group was genetically related to modern Caucasoid populations and flourished in the late Paleolithic culture. The Iberomarusians' range "extended from Spain, to Morocco, Algeria and Tunisia."[40] This group generally lived within 100 km of the Mediterranean, and were hunters of Barbary sheep.[41] Rock engravings found in the Sahara are dated as early as 9,000 B.C.E, the earliest depicting the exploits of a hunter-gatherer class.[42] These engravings were followed by paintings, dated around 3,500 B.C.E., which showed a new group of humans with both Caucasoid and Negroid faces, who had clearly domesticated a type of cattle now not known.[43] The cattle depicted are hypothesized to have arrived on the scene as early as 7,000 B.C.E.[44] When the Iberomarusians disappeared around 7,500 B.C.E., they were replaced by the *Capsians*, possibly of local origin, who were hunter-gatherers and fishermen, and consumed large quantities of mollusks.[45] The Caspians became domesticators of sheep and other animals, acquired pottery, and painted on ostrich shells.[46]

One of the many commonalties that the ancient Spaniards and Moroccans share is a preponderance of historians who wish to attribute most of their accomplishments to outsiders. Nonetheless, as North African historians note, those who indigenously dwelled in what we now call Morocco did not skip either the Copper or the Bronze Ages, though the actual materials may have been supplied by the Iberians.[47] Like their northern neighbors, the indigenous (*Imazighn*) peoples of what would become Morocco cultivated wheat and raised cattle prior to the arrival of the Phoenicians.[48]

The religious practices of the Imazighn are open to some degree of debate. However, evidence collected from Imazighn tombs paints a fascinating portrait of a people with a strong belief in an afterlife, a possible pantheon of gods, and the belief of Divine intercession through the medium of dreams.[49] According to Brett and Fentress, the dead of the Imazighn were believed to be directly "connected to the fertility of the soil and probably exercised some control over the future."[50] Some, such as the Augilae tribe, believed that the spirits of their ancestors were gods, and "[swore] by these and consult[ed] them as oracles, and, having made their requests, treat[ed] the dreams of those who [slept] in their tombs as responses."[51] Such a conclusion has been drawn from the fact that archaeologists have encountered numerous Imazighn tombs with antechambers for the living, so individuals could spend the night and hopefully have dreams influenced by the deceased.[52] Such funerary worship was often mixed with iconographic references to fertility, not entirely unlike the tradition that has lasted into modern times wherein Moroccan women visit the tombs of salihs (those held to be saints) in the hope of receiving *baraka* and becoming pregnant. Clearly, while ancient Imazighn practices had their own integrity, one could make the argument that they have been subsumed into what we now call Moroccan Islam; a form of worship that arguably includes elements which predate the life of the Prophet.

One can thus conclude that the Maghreb, and specifically its far western corner that we now call Morocco, was a place of convergence for different cultural and ethnic groups in prehistorical times. Humans who migrated from Sub-Saharan West and East Africa and from Europe ended up living in close proximity in Morocco. As L. Luca Cavalli-Sforza has noted, by the year 1,000 B.C.E the rock paintings began to depict a new animal, the horse, which was resistant to dry heat and was most likely brought from the Middle East or Greece to drive chariots.[53] The arrival of the horse was followed by the desertification of the Sahara. By the first millennium B.C.E the horse could no longer cope in the progressively drier climate of the Sahara, and was thus replaced by the camel, which arrived in Egypt from Asia in 1,600 B.C.E and eventually made its way west via human migration.[54]

The crops cultivated by the Imazighn[55] included breadwheat and barley, and there is palaeobotanical evidence that the Imazighn cultivated marsh species of plants, which the current dryer climate can no longer support.[56] Alongside these crops there remained a domesticated version of sheep, an animal the Imazighn are dependent upon to this day.[57] However, by the first millennium B.C.E, archaeologists have come to know the most about the Imazighn not by their crops but by the elaborate tombs they left behind.[58]

Morocco and Spain as a Lingual and Cultural Crossroads

While modern observers of Morocco and Spain will be the first to emphasize the economic, geostrategic, and cultural bridge these two countries engender across the North–South split, fewer linguistic scholars of ancient cultures are willing to make such hard-and-fast connections between these two nations. Reasons for this phenomenon are varied, and might include the following: (1) an overvaluing of political or continental differences between the two nations; (2) a great investment in separating European and African origins in the name of simplifying history; (3) a lack of substantive speculation as to who first crossed the Strait of Gibraltar; and (4) an inherent racism on the part of those who would keep European origins "pure" to satisfy their own cultural and political assumptions, fears, or agendas.

As we have already noted, geneticists readily admit that early Caucasoid humans lived on both sides of the Strait of Gibraltar.[59] Further, we have seen through the work of art historians that cave paintings found in Morocco depicted both Caucasian and Negroid faces among their figures.[60] In addition to this evidence for the human porousness between these two regions, linguists present yet another piece of important evidence.[61] Long a hotly contested topic, some linguists have argued that there is an Imazighn root to some of the Iberian languages, particularly Basque.[62] The linguist Jorge Alonso posits that Basque and other prehistoric and "mostly forgotten languages" were brought to Europe from North Africa.[63] There are admittedly problems with these theories, especially because Basque is considered to be a northern Spanish language, and thus the Imazighn would have had to skip over the *Tartessians* in the south of Spain to reach what is now known as the Basque region (a region that may well have been much larger in earlier eras). Nonetheless, linguists other than Alonso have posited the same theory, particularly Hugo Schuchardt, James M. Anderson, and Hans Mukarovsky.[64] It is interesting to note that just as the northern Imazighn were thought to be mountain dwellers, so too are the Basques. Antonio Tovar is also a linguist who is an advocate of the North African–Basque connection, though his methodology, like many of those advocating such a position, has come in for heavy criticism by his contemporaries. Such criticisms are based on the grounds of too small a word sampling and possible connections based on borrowings from Latin-Romance rather than Imazighn languages, among other points of contention.[65] Like any discipline, however, linguistics is always undergoing revisionary scholarship. As the circle of those allowed to participate in the discussions regarding the Basque–Imazighn lingual connection widens beyond Europeans and Americans, different theories will no doubt come to the fore.

Thus through art, genetics, and linguistics, we can see that what might first appear to be an affirmation of the North–South split is in fact a clear questioning of its numerous assumptions. If the Iberomarusians lived on both sides of the Strait in ancient times, they can thus be arguably seen as one people. It is therefore only retrospectively that we have become comfortable with imposing the parameters of what we refer to as the modern North–South split upon vast tracts of history, and thus we are now called to critically question many of the assumptions and parameters of how we have come to divide Africa from the European landmass. In this vein, it should be noted that many historians of the time of the Greeks and Phoenicians through the era of Augustine point out that the Mediterranean was long viewed as a whole, connecting rather than dividing people from one another.

Level Two: Conquerors and Crops, Rulers and Religions—From the Phoenicians to the Romans

As the historian Abdallah Laroui has noted, knowledge of the long period of Moroccan history that begins in the second millennium B.C.E and ends in the seventeenth century is limited in that it is only "known to us through Greco-Latin literature."[66] Much of early Spanish history is also a product of the recordings of its conquerors. Among the first of these to colonize both Iberia and the Maghreb were the Phoenicians, whose origins are thought to be in the eastern Mediterranean (possibly in what we now call modern Syria). An early form of venture capitalists, the Phoenicians' drive to explore and conquer is thought to have been rooted in their desire to acquire more raw materials and to sell their products beyond the borders of their homeland. As the first to circumnavigate the African continent, the Phoenicians were as mobile as they were resourceful. They founded many colonies on all sides of the Mediterranean around or before the eighth century B.C.E., including numerous settlements in both Morocco and Spain. The remains of their greatest urban achievement, Carthage, are found in modern-day Tunisia.[67] One of the principal natural resources sought by the Phoenicians was the shell of the Murex, the source of a purple dye used in royal and aristocratic robes.

It is difficult to speak definitively on the subject of Phoenician religion because of the lack of primary source liturgical and mythological texts. However, one can infer important information from Phylo of Byblos' first- or second-century C.E. document, *Phoenician History*.[68] We know, for example, that the Phoenicians were likely polytheists, and "venerated a plurality of super-human beings, . . . who represent as a whole the totality of [humanity's] and society's interests and needs."[69] The Phoenicians believed that they received "benefits, favors and protections from these gods."[70] For the Phoenicians, it was often the female deities that held the preeminent position in the pantheon of gods.[71] These goddesses were most likely connected to fertility, prosperity, love, and war (much like the Greco-Roman goddesses that would follow).[72] Scholars, however, have not been able to agree if the importance of the female deities translated into a higher status for human females.[73] We do know, however, that the Phoenicians attached particular spiritual significance to specific geographic locales. Many Phoenicians worship practices were tied to the cycles of the land and the heavens. As Glen Markoe has recounted:

> The cultic calendars of the various Phoenician cities were governed by a
> prescribed series of feasts and celebrations that revolved around the

agricultural cycle. Sacrifices were offered in celebration of the New Year and the advent of plowing and harvesting; as various texts suggest, solar and lunar worship played a prominent role in the Phoenician calendar, which was calculated on observation of the new moon. Integral to the cycle was the spring awakening or resurrection of various vegetation deities, such as Melqart of Tyre and Eshmun of Sidon. In the Tyrian Melqart festival . . . the god was burned in effigy on a ritual pyre, and later resurrected through a ritual marriage with his spouse, Asarte. The ritual celebrated not only the cyclical rebirth of nature, but the restoration of the cosmic order over which Melqart and the Tyrian king ruled.[74]

As the Phoenicians traveled, they encountered new deities specific to particular communities. Thus, each community had its own unique religious identity.[75]

The influence of the *Jews* of the diaspora, however, persisted long before and long after many different visitors and conquerors had passed through the Maghreb. From the longest continually inhabited settlement in Ifrane (fourth century B.C.E.) to the significant exodus to Israel of the twentieth century, the Jews have played a critical role in the cultural, religious, and economic life of the Imazighn and the Maghreb's many subsequent conquerors.[76] Much of what we now know of the Jewish influence on the Imazighn comes to us through twelfth-century C.E. writings, and later from the writings of Ibn Khaldun, who recounted "the story of the Jewish 'priestess' Kahina, who led the resistance to the Arab conquests of the 7th century CE."[77] Some traditions passed down through these writings refer to Imazighn Jewish tribes, though the exact time of their conversion is contested. Nonetheless, it is clear that multiple Jewish communities were established during the time of the Romans in response to the destruction of the second temple (70 C.E.).[78] Later, as Rabbinical Judaism established itself and spread, it ultimately came to be in competition for converts among the Imazighn with Romanized Christians.[79] Thus the Imazighn came to bear the religious imprint of the first two Abrahamic traditions prior to the arrival of Islam in the seventh-century C.E.[80]

The *Greeks*, who arrived in the western Mediterranean during the Phoenician period, were known to have planted olives wherever they traveled, as well as grain on many occasions. Other typical Greek crops that were likely brought to Iberia and the Maghreb included figs, barley, grapes, wheat, and hay.[81] Like the Phoenicians, the Greeks established a number of colonies at both the eastern and western ends of the Mediterranean, including Iberia and the western Maghreb.

Regarding the religious beliefs and practices of the early Greeks, J. Donald Hughes has made some compelling observations:

The attitudes of the early Greeks toward nature were shaped by their religion. They saw the natural environment as the sphere of the activity of the gods. Greek religion was in large part the worship of nature, and the old Greek gods were essentially nature deities. The gods ruled nature, they appeared in it, they acted through it; therefore, human activities which affect the environment often were seen as involving the interest and reaction of the gods.[82]

Like the Phoenicians, the Greeks were also polytheists, who laid the groundwork for what was to become the foundation of Roman religious belief. For the Greeks, the human was beholden to the gods, but was not without the ability to act in order to affect one's fate. According to J. Donald Hughes, "Greek religion had a strong sense of its natural locality."[83] In this respect, Zeus can be seen as a sky God (some call him a weather God, reflecting his concrete affect upon the earth) while Poseidon controlled the waters of the seas. The Greek's shrines and temples were built in a manner that was highly attuned to their natural settings. "Altars and places of worship were originally in groves of trees."[84] Despite such beliefs, however, the Greeks violated their own spiritual laws regarding the treatment of the natural world on many occasions, and in so doing they expected to incur the wrath of the gods through acts of natural disaster.[85]

This can be seen in the fact that the Greeks, having devastated their own native forests, and many of the creatures who depended upon them, were forced to seek wood from outside of their territorial waters.[86] The Greeks were also prolific hunters, deeply cutting back the numbers of native animals in their own territory before setting out to do the same outside of Greece. These Greek animals included the beaver, copious numbers of birds, lions, wolves, wild cattle, sheep, and goats.[87]

While the Phoenicians and, to a lesser degree, the Greeks had sought precious metals in Iberia and the Maghreb, the *Romans'* arrival in Iberia (207 B.C.E.) hailed the introduction of a people who came with the intention to denude the land of trees and supplant the former forests with wheat. In addition, the Romans dug deeply into the mineral reserves they encountered, with long-term devastating effects.[88] The Romans, as they had in Syria, also engaged in a most crippling form of deforestation in what is now Morocco.[89] Maghrebi marshlands, including coastal forests in particular, were eliminated many kilometers inland under the Romans, in which they sought to cultivate wheat instead. The Romans used elaborate drainage techniques in these former marshlands, in some cases pushing the coastline significantly further out to sea.[90] Roman deforestation led to the tremendous soil erosion, which washed deposits of rocks, sand, and mud into the Mediterranean.[91] This problem was exacerbated by the Roman propensity to overgraze their domestic livestock, a practice they began in their own homeland.[92] While the Greeks had been avid hunters, the Romans outstripped Greek consumption of wildlife in almost every category. The Romans used animals for their fur, ivory, their plumage, and their usefulness in the coliseum, the single most destructive Roman pastime of all.[93] In the case of the Romans, hunting was not limited to the upper classes, but was in fact a tradition enjoyed by the peasantry as well.[94] Under the Romans, the elephant, rhinoceros, and zebra became extinct in

North Africa, a fact that the Romans boasted of, claiming that they were "removing dangers to [humans] and to [their] agriculture."[95] At the same time, however, it was the Romans who introduced some of the first Northern attempts at conservancy, declaring particular forests sacred and therefore preserved, planting trees in lumbered areas, and creating artificial parks that became a protected domain.[96] As resources dwindled the Romans created tree plantations (documented in Egypt), which were regulated, prohibiting entry to animals such as goats or sheep, and governed "by laws stipulating the age and size of trees that were available for logging or thinning."[97]

While the Romans inherited the same pantheon of gods from the Greeks, their beliefs regarding nature were far more utilitarian. The Romans held that the Earth existed to meet human needs, and they approached their task of empire building with an overarching belief in order and their own superior position within that order. In this respect, the Romans saw nature's task as that of conforming to human needs, which cemented a new approach to development in the ancient world.[98] In this respect, Rome set the standard that is still followed by many in the twenty-first century and therefore represented a profound theological paradigm shift, to the celebration of a theological anthropology which places the human as the descendent of God on earth.

Roman religion was not without its ties to and respect for nature, however. The Romans held a reverence for wild places, yet their ties to nature were more agricultural than "natural."[99] This tendency is reflected in the fact that the Roman calendar was based upon the agricultural year.[100] Like the Greeks, the Romans had their own weather gods. Jupiter was the bringer of rain, for example. Yet as the Roman Empire expanded, so too did its pantheon of gods, to include the household gods of the conqueror and the conquered. The Roman religion was thus highly syncretistic, and succeeded for a time in supplanting the former religions of those colonized. When the Roman Empire became Christian under the consolidated rule of Constantine (324 B.C.E.), the Roman Empire pursued the Christianization of its subjects with the same fervor it had proselytized its former religious traditions.

According to J. Donald Hughes, one can view the ultimate decline of Roman rule as a result of its ecological unsustainability.[101]

Of course, the fall of Rome was a large and complex phenomenon that cannot be attributed to a single cause. Ecological failures interacted with social, political, and economic forces to assure that the vast entity called the Roman Empire would disappear or be changed beyond recognition... "When Rome falls, the world falls, too," ran an old saying quoted by Gibbon and Byron, and it may be time to consider the converse, for when Rome made nature a slave and tried to work that slave beyond endurance, the natural world "fell," or at least lost the ability to support the mistress of the world, and Rome fell too.[102]

Yet as we shall soon see, while Rome pioneered many of the destructive ecological consumption techniques we know and practice today, the destruction they wrought did not simply reappear in the nineteenth or twentieth centuries. Rather, they were consistently nurtured by a successive group of foreign and domestic occupants of Iberia and the Maghreb.

From the Vandals to the Byzantines

After more than five centuries of Roman rule, Iberia and the coastal territories of the Maghreb came under the control of competing branches of the *Vandals* in the fall of 409; finding the *Seubi* in the northwest or Galicia, the *Silings* in the south in Baetica, and the *Alans* on the northern coast of the Maghreb.[103] Not to be excluded, a branch of the *Visigoths*, the *Athaulf-Goths*, invaded Catalonia in 415.[104] These invasions ushered in a time of great tumult and bloodshed, which stood in stark contrast to the latter relatively peaceful rule of the Romans. The Athaulf-Goths, employed by the Romans to extricate the Vandals from Spain, came close to success, before the majority were withdrawn from Iberia by Rome in 418.[105] While in Spain, the Visigoths' main center of power was located in Castile and Segovia, where one particular group cultivated corn, and many, numbering in the thousands, remained present until 711.[106] The Visigoths were horsemen and ranch tenders, who viewed themselves not as socially independent, but rather as a supplanted extension of Roman culture and custom.[107]

Historians are not certain whether the Vandals converted to Arian Christianity before or after having arrived in Spain.[108] It is agreed, however, that the Vandals' conversion to Christianity was a clear attempt to find a path toward sharing in the benefits of the Roman civilization.[109] However, despite their desire to capitalize on Roman advantages, the Vandals chose to theologically oppose Rome itself. The so-called Arian Heresy (according to Rome) to which the Vandals subscribed, held that "Logos [was] a creature called into being by God 'out of non-existence.' "[110] The Arians also believed that there could be no mediator between God and creatures, and thus the Son of the Trinity was in fact a creature.[111] Such Trinitarian debates had begun nearly a century prior to the Vandal's arrival in Iberia, and they prevented Contantine's version of Christianity from prevailing throughout the Roman Empire. The Vandals attempted to solidify their hold over Iberia and the Maghreb through marriage alliances with prominent Romans. At the same time, the Vandals sought to capitalize on the pre-Christian Roman tradition of Emperor worship and to draw such veneration toward their own provincial leaders in the Maghreb.[112] It has been hypothesized that such a form of Emperor worship was in fact made secular by the Vandals due to their acceptance of Christianity.[113] Thus the Vandals attempted to perpetuate Roman institutions, while negating the original religious undergirding of tributes and traditions established before the conversion of Constantine.

The Vandals proceeded to consolidate their positions in Iberia and ultimately, in 429, crossed the Strait into northern Maghreb. Eventually, a Visigothic and Vandal alliance was spurred by what became their common enemy, Rome. Agriculturally speaking, it was during this period that the Vandals sought to consolidate control over many lands that had been providing Rome with grain.[114] By 439, the Vandals controlled all of what had been Roman Africa, including Mauretania, Numidia, Cyrenaica, and Aegyptus.[115] During this period, it is speculated that the Vandals attempted to perpetuate Roman agricultural practices, while beginning to blur the distinction between those who had been called Goths and Vandals. Together, their territory embraced much of Iberia and present-day northern Morocco with an unbroken human chain.

The Imazighn's interactions with the Phoenicians, Greeks, Romans, and the Vandals greatly impacted their worldviews, their agricultural modes of production, and their manner of politically organizing themselves. Having been colonized by the Carthaginians, the Imazighn emerged from their experiences emulating those who had controlled them. From the Phoenicians it is hypothesized that the Imazighn adopted the use of the title of king, and subsequently consolidated larger territories of control under this new form of leadership.[116] The Greeks' contribution, among others, was to bring their language to the northern urban centers of the Maghreb. The Imazighn, meanwhile, attempted to organize their territories, modeling their own more inland kingdoms after the Romans who had departed. The Romans also brought Christianity and the language of Latin to the Maghreb, especially in Ifriqia, greatly impacting Imazighn cultural and religious practices. By the twelfth century, however, the Muslim conquest erased most of the vestiges of Christian practice among the Imazighn of the Maghreb. At the same time, the order that Roman customs brought to Imazighn political practices was undermined by the arrival of the Vandals, after which it is thought that the Imazighn leaders found their territorial populations much more difficult to control in a centralized manner.[117]

The Visigoths, like the Vandals before them, embraced an Arian form of Christianity. As the Visigoths invaded Iberia, they took for themselves the land previously held by Romans, while spreading a form of Christianity that was anathema to the land's former occupants. The Visigoths had become Christians in the fourth-century C.E. led by their king, Fritigern, who converted in 376.[118] Prior to their conversion we find among the Visigoths (and likely the Vandals) a focus on tribalism, a pantheon of deities, and rituals of sacrifice and worship.[119] Some scholars, such as Andreas Schwartz, have posited that prior to accepting Christianity, the Visigoths held sacrifices and feasts in conjunction with the cycles of the moon.[120] The Visigoths were also among those Germanic tribes that celebrated the winter solstice, further embedding their religious practices in the turning of events of the nonhuman world. Ancestors, gods of war, and specific rivers are also believed to have been worshipped by the

Visigoths.[121] Such phenomena beg the question: to what degree did the Visigoths' original forms of nature worship impact the manner in which they accepted and practiced Christianity?

By the middle of the sixth century, the Visigoths had expanded a nominal control to southern Spain, only to find themselves confronting a Byzantine invasion of the same territory. The Byzantines, whose forces remained until 629,[122] were known for taking an extraordinary amount of the local crops for themselves.[123] In response, the Visigoths returned to central Spain, and remained until the Muslim invasion of 711.

Like the Arians before them, the Byzantines were in direct theological conflict with the Western, Latin-speaking Roman Church. With the Byzantines came the practice of the veneration of icons, and the steadfast belief in the "full and distinct human nature of Christ... "[124] In another break with the West, the Byzantines also held to the doctrine that stated that Christ was of two wills; a premise supported by Jesus's praying to God in the garden of Gethsemane, "Not my will but thine be done."[125] The Byzantine's use of sacred images rankled the Western Romanized branch of the church, which finally decreed under Leo III in 730 that such images were not to be tolerated (despite the fact that images would later play a central role in Romanized Christian worship).[126]

The Muslims and the Spaniards

There is much speculation as to the exact circumstances that ushered a Muslim army across the Straits of Gibraltar in 711. Some believe that it was at the invitation of a displaced Visigothic contender to a southern throne. Others would explain the military action in terms of a simple continuation west of the spread of Islam, which had begun less than a century before at the death of the Prophet in 632. Islam, having arrived from the east in what is now Morocco by way of the troops led by "Uqba ibn Nafi," the governor of Kairouan, displaced the last vestiges of the Byzantine forces at the northern port of Sebta in 680.[127] What is clear is this: the Imazighn Tariq Ibn Ziyad, governor of Tangier and under the tutelage of the Damascus-based Umayyad Dynasty Arab Musa Ibn Nusayr, led 7,000 Imazighn troops, newly converted to Sunni Islam, across the Strait and defeated the Christian army led by Roderick.[128] They were soon joined by 18,000 Arab troops led by Musa.[129] Within ten years, those who came to be known as "*Moros*" (in English: "Moors") to their European hosts had gained control over nearly the entire Iberian Peninsula, save for the northern mountains of Asturias.

The first brand of Islam to enter the western Maghreb profited greatly from the prior presence of Christianity, which had already sown among the people an acceptance of monotheism by the Romans and the Byzantines. The middle Maghreb (modern Algeria) offered a Christianity that in many cases had developed independent of the church, and which had evolved into what Abdallah Laroui calls "an abstract monotheism capable of accommodating itself to any dogma whatsoever."[130] It was in this part of Maghreb, which had been independent for two centuries, that the "invading Arabs first attempted to [spread] their new religion."[131] Still, the act of conquering North Africa was hardly easy; in fact in required over 50 years of fighting both the Byzantines and the Imazighn. Under the leadership of Umayyad-backed Musa, the brand of Sunni Islam brought by the Arabs came from their original base of operations in Damascus, via Kairouan in modern-day-Tunisia. The type of Sunni Islam that arrived on what would be Moroccan soil was not theologically aggressive, and was further tempered by the belief that the forced conversion of the "People of the Book," Jews and Christians, was forbidden. The Umayyads also brought with them the theological battles that had originated in the Levant; particularly the countering claims of the Kharijiri, who cleaved to a stricter interpretation and focus upon the Qur'an, a resentment of the wealth of the cities, and an ardent form of piety that pitted them against mainstream Sunnis.[132] Thus the focal point of the so-called Islamization of western Maghreb was on the Imazighn.[133] Therefore we can understand why it was the Imazighn who joined the original push by the Umayyads into Spain. In time, however, people began to travel both east and west, a phenomenon finding Maghrebis studying Islam in the Levant and involving themselves in the same theological debates from which the original Arab invaders had emerged. In time, North Africa was to be embroiled over the promises held out by the orthodox Mu'tazilism and juridically based Hanafism of Baghdad, over against the ultimately prevailing Maliki school of law and its central North African text, the Mudawanna. The Mudawanna was one of the most well-respected collections of Hadith in all the Muslim world, containing the traditions of the Prophet, legal decisions of early Muslim jurists, and reports of the Prophet's companions.[134] Thus, the Maliki school of law is a juridical form of Islam, whose founders' writings remain central to modern Muslim legal scholarship. The Maliki school of law stressed piety and asceticism, in what was an orthodox interpretation of Sunni Islam. The Maliki school of law was to hold sway in Morocco during its formative years as an Islamic region.

While newly converted Imazighn and newly arrived Arabs clashed over competing interpretations of Islam on the southern side of the Strait, those indigenous North Africans, joined by varying Levantine forces, consolidated their control of the majority of the Iberian Peninsula. This was a period that ushered in a profound degree of religious pluralism. While some Imazighn rejected Islam outright, others adopted the Kariji interpretation of Islam, a "rigorously Qur'anically based movement, fiercely egalitarian and dedicated to divine justice."[135] It was those who sided with the Kariji approach to Islam who fought (both doctrinally and militarily) against the ultimately prevailing Umayyad and Sunni-identified forces. Still other Imazighn sided with the "Barghwata" movement of Islam, an "indigenized" form of Islam that included its own Qur'an in Tamazight as a counter to the Arabic Qur'an of those who came to be called the

Umayyad-alligned Sunni.[136] This provides a partial explanation as to why, simultaneous to the Muslim conquest of the bulk of Iberia eventually known as "Al-Andalous," Arab forces pressing into the interior of Morocco were met with stiff resistance by multiple rural Imazighn populations.[137] For this reason, many Arab forces focused on urban centers in their efforts to pacify the population.[138] Abdallah Laroui writes that after 711, what we now call Morocco "became theoretically a province of the Arab empire, providing soldiers and slaves, and paying tribute to fill the coffers of the Caliph of Damascus."[139] One can only assume that such tributes included agricultural products as well as, cattle, sheep, and coinage.

As time progressed, wave after wave of Arab and Imazighn dynasties came to rule portions of Morocco and Iberia, as well (at times) completely controlling both. Yet within these same dynasties coexisted Visigothic Christians, Roman Catholics, Sunni, Shi'i, Sufi, and hybrid Muslims, and a rich and thriving Jewish culture. This was an extraordinary time of cultural syncretism, especially in Iberia (yet certainly in the Maghreb as well). As the historian Juan Zozaya has written:

> The entry of the Muslims into the Iberian Peninsula meant the establish-
> ment of a cultural syncretism made up of Byzantine, late Roman, and diverse
> Hispanic elements, as well as of Arab, Berber [Imazighn] Syro-Palestinian,
> and Islamic Mesopotamian elements, to which would be added from the
> beginning of the tenth century onward, especially Chinese and Persian.[140]

The Damascus-based Arab/Sunni *Umayyads* (661–750 C.E.) were ultimately relegated to Iberia to be replaced in the greater Maghreb by the Arab/ Shi'i-Sunni *'Abbasids* (750–788). The 'Abbasids, who came to power in the east with Baghdad as their capital, were originally inspired by the Shi'i in rebelling against the Umayyads, and only later became institutionalized as Sunni. Eventually, the 'Abbasids were conquered by the Arab–Imazighn/Shi'i *Idrissis* (788–974), who originally fled the Abbasid-controlled east to take refuge in Morocco, and were the first to unite Morocco under one head.[141] It was the Idriss I who originated the idea of integrating Imazighn blood into the royal house via the women he married. The Shi'i/Arab *Fatmids* (910–1073) usurped the Idrissis and controlled Morocco in the tenth and eleventh centuries. While in many cases what occurred in Morocco was reflected in Iberia, in many cases it was not. The Umayyads, for example, held sway in al-Andalous long after their power had waned in the Maghreb. This mixture of competing dynasties led to sometimes bewildering results; between 716 and 746 C.E., for example, the city of Cordoba was led by a total of 19 different governors.[142] The dynasties that followed the Fatmids are as follows: the Imazighn/Sunni *Almoravids* (1060–1167); the Imazighn/Sunni *Almohads* (1133–1269); the Imazighn/Sunni *Marinids* (1244–1465); the Imazighn/Sunni *Wattassids* (1472–1554), followed by the Arab/Sunni *Sa'adi* (1554–1659); and finally the current Arab–Imazighn/Sunni ruling family: the *Alawis* (1668–present) who have led Morocco for over four centuries.[143] With each subsequent dynasty came different levels of tolerance for the religious and cultural diversity among the peoples of al-Andalous, as well as in the western Maghreb. Many were forced to flee south or north of the Strait for their own safety, to convert, or to feign conversion, depending upon the current rulers, their own personal circumstances, and their ethnic and religious affiliations.[144]

In the midst of the Islamic movements and their representatives who washed over the Maghreb and al-Andalous, it is impossible to ignore the critical contribution made by Sufi branches of Islam. While some scholars would place Sufism as a later development following the Sunni and Shi'i movements, others hold that Sufism is in fact the essence of Islam upon which all later Islamic movements were built.[145] As Frederick Denny notes, "the major roots of Sufism are indigenous to Islam. These are the Qur'an and the Prophet Muhammad."[146] In Sufism, it is sometimes difficult to draw a clear distinction between the Qur'an, the Sufi's principal source of inspiration, and the Prophet.[147] This is because for many Sufis, the role of Muhammad is that of the "universal human, the model for authentic existence in union with God."[148] Thus, Hadith is central to many Sufi practices, yet the Sufi's collection of Hadith includes accounts that are not found in so-called traditionalist Sunni and Shi'i versions. While the so-called mainstream expressions of Islam are drawn to searching for the "literal" meaning of the Qur'an, known as *tafsir*, the Sufis are generally drawn toward the allegorical and symbolic interpretation of the Qur'an, known as *ta'wil*.[149] Denny notes, "the Sufis emphasize God's love more than his justice, without, of course, denying the latter in any sense."[150] In this love, one finds a quality of mutuality between believers and God that is not readily found in other expressions of Islam.

Unfortunately (and many would say "unfairly"), for many "mainstream" Muslims, Sufism is viewed negatively for its emphasis on the mystical, its sometime focus on deep asceticism, and some branches' embrace of what is viewed as ecstatic worship practices. Yet it is impossible to generalize about all those who seek the Sufi way, or *tariqa*. There are many branches of Sufi religious expression, and many "brotherhoods" who (together or apart) seek the Sufi way of life. Sufism, like all the other Muslim movements that arrived in the Maghreb and al-Andalous, has its roots in leaders and followers from the Levant. While an anti-Sufi Maliki school of law of Sunni Islam originally held sway in the Morocco, it ultimately converged with the Sufis to construct what would become the dominant expression of Islam in the western Maghreb.[151] With its base in Marrakech, this expression of Sunni–Sufism emerged during the Almohad Dynasty, providing what Abdallah Laroui calls the first "truly popular ideology" among the Moroccan people.[152] It was during this period that Marrakech achieved an importance in the Muslim world that rivaled Baghdad, Cordoba, and Cairo. It is thus impossible to explicate Moroccan Islam in the absence of Sufism, as it embraced and was embraced by many among the rural and urban populations of Morocco, with the attendant amount of syncretism one might expect of any successful insertion of an originally "foreign" religious practice.

The impact upon agricultural practices under the Muslims was unprecedented and extraordinary. The Muslims living in Spain were renowned for their innovations. Many agricultural practices originated in the Levant, but it was in al-Andalous, with its rich land and more plentiful rain that many techniques came into their own. At the same time, one must note that as the Muslim empire grew, so too did its agricultural knowledge. Many vegetables now known as "Mediterranean" were in fact brought by the Muslims from as far away as Asia, heralding an unprecedented amplification of agricultural possibilities. According to Norman Roth, the Muslims not only managed "to grow every conceivable type of fruit and vegetable known today, but certain crops (such as rice and cotton) were produced where they had no reason to be expected to grow."[153] In addition, the Muslims are said to have perfected horticultural techniques that allowed them to inject syrups and perfumes into trees to produce a sweeter type of fruit.[154] Even more extraordinary, notes Roth, the Muslims "were capable of producing blue roses and growing roses (and apples) out of season, [and even] of developing letters and pictures on growing fruit, and other wonders."[155]

More importantly than changing the color of a rose, however, was the Muslims' introduction of three-crop rotation.[156] No doubt, such a feat was aided by the fact that the Muslims had improved immensely on the existing aqueduct systems first built by the Romans, which enabled farmers to receive a more consistent and plentiful source of water in areas previously unarable. The Muslims, having previously been forced to work with substantially less water in their own lands to agriculturally support their populations, employed technologies that offered the least amount of waste and the most productive output with regard to water use. Such agricultural advances were not kept a secret by the Muslims, as there is ample evidence that Muslim agricultural advances influenced Christian Visigothic practices as well.[157] Thus, under the Muslim rule, the area of arable lands both in Iberia and the Maghreb increased substantially for both Muslim and Christian landholders. Under the Muslims as well, the production of olive oil for trade was expanded, as was (in al-Andalous) the production of grapes. At the same time, as communities became more sedentary on both sides of the Straits, long-term concerns regarding the viability of grazing lands for cattle and sheep must have come to fore.

This same period saw the extraordinary development of a time of peace and relative harmony among the faith traditions of Islam, Judaism, and Christianity on the bulk of the Iberian peninsula, which came to be known as the Convivencia. Never before on European territory had all three of the Abrahamic traditions coexisted in such a manner. While it is true that some Muslim dynasties lent more than others to a constructive cooperation among the faith traditions, the Convivencia that did exist stood in stark contrast to the society formed in northern Iberia by the Europeans ultimately linked to the Inquisition.

This period of time was also marked by the introduction of what Bernard Riley calls "plow culture" and "transhumance stock raising at the expense of the older hoe culture" to those people who would one day be called the "Northern Spanish."[158] Techniques brought north by Christians who had lived under Muslim rule greatly expanded the scope and quality of northern Iberian agriculture under the Europeans.[159] Out of these technological advances "emerged a dense, peasant agricultural world [in the Northern region] . . . probably for the first time in its history."[160] These improvements allowed the European northerners to economically recover from Muslim incursions, and laid the monetary and social groundwork for what would ultimately become the Spanish push south.[161] For as agricultural improvements altered the lifestyles of at least some of the more prosperous Europeans in the north, they also created "more complex governmental and social institutions which were soon to enable [Iberia's northern Europeans] to project [their] own strength against a Muslim al-Andalous in growing disorder."[162]

By the 1230s the Christian reconquest (or "Reconqista," as it came to be known among the group we later refer to as "the Spaniards") had descended from the north of Iberia, which modern Spanish historians claim originated in Asturias in 722, and later came to embrace the kingdoms of Aragon, Castille, and Leon. This coalition of often interwarring kingdoms ultimately united in their efforts to recapture Iberia from the Muslims, forming a "crusade" against those they viewed as the peninsula's "heretical occupiers." Yet despite Spain's historical claim to having first united the Iberian Peninsula, there is a different reading of the same story by North Africans. This is because ironically, one could argue that Spain as a nation did not really come into existence until the majority of it was united by the Muslims, the Iberian Peninsula previously serving as simply a battleground of smaller kingdoms and tribes.

While some Christians known as the Muslim-influenced Mozarabs had found a way to successfully coexist with their Muslim rulers, the Christian forces that descended from the north were not interested in theological or geographic compromises of any kind. In the midst of these changes, the church affiliations of many Europeans changed as well. While the Mozarabs and their unique liturgies had been tolerated before, the region under Alfonso turned its attention to Romanizing all churches under its control.[163] This act laid the crucial groundwork for what was to become the eventual uniting of those who sought to capture the Iberian Peninsula, under the banner of a very aggressive and nonapologetic form of Roman Catholicism undergirded by the Inquisition.

By 1246 Christian armies had entered cities as far south as Cordoba, and in 1258 Seville fell. Some have argued that the Christian victories were aided by the fact that Muslim hegemony was mostly limited to urban areas, leaving the Mozarabs control of great swaths of the countryside.[164] At the end of the thirteenth century, the bulk of Islamic-controlled Iberia had been pared down to the Caliphate of Granada, which managed to survive two centuries more through skillful diplomacy and tributes. Meanwhile, the rest of Iberia was divided into five territories: Navarra, Aragon, Castilla, Leon, and Portugal, entities that were often at odds with each other. However, the unrest caused by the rivalries among these regions was matched if not surpassed by the first arrival of the Plague in Iberia (1348–1351), a time that coincided with the first Christian-sponsored anti-Jewish pogroms on the part of the advancing northern Europeans.

In 1492 the Muslim reign in Iberia came to a close when Ferdinand of Aragon and Isabella of Castile, having earlier united under the banner of the Roman Catholic Church and the Inquisition, entered Granada and forced the Caliph to publically submit to their authority, effectively bringing over seven centuries of Muslim rule on the peninsula to an end. Both the Muslim and Jewish communities were given a stark choice under Ferdinand and Isabella: convert to Christianity or face death or expulsion from the Iberian Peninsula.[165] Thus, a tripartite accomplishment on the part of the northern Europeans was brought fully to the Iberian Peninsula: the reconquest and eventual consolidation of all of what would come to be known as Spanish land occupied by Spanish people; the expulsion of the Muslims; and the expulsion of the Jews. Arguably, one could add to this list of accomplishments the construction of the modern Spanish identity itself, which arguably did not exist prior to the Muslim unification of the bulk of Iberia and their subsequent demise. All these events occurred under the auspices of the Inquisition, which offered the original rallying cry that had inspired many Christians to join the Reconquista on religious grounds. The spirit of this period is brilliantly captured in the logbook of Cristobal Colon, better known among English speakers as Christopher Columbus, whose recorded writings that document his first voyage across the Atlantic actually begin in Granada:

> On January 2 in the year 1492, when your Highnesses had concluded their war with the Moors who reigned in Europe, I saw your Highnesses' banners victoriously raised on the towers of the Alhambra, the citadel of the city, and the Moorish King come out of the city gates and kiss the hands of your Highnesses, and the prince, my Lord. And later in the same month . . . .your Highnesses decided to send me, Christopher Columbus, to see those parts of India and the princes and the peoples of those lands, and consider the best means for their conversion . . . . Therefore having expelled all the Jews from your domains in the same month of January, you Highnesses commanded me to go with an adequate fleet to these parts of India . . . I departed from the city of Granada on Saturday May 12 and went to the port of Palos, where I prepared three ships.[166]

At this point, it is critical to note that while anti-Muslim sentiment was effectively used to unite otherwise warring European factions, the end of the Reconquista is arguably the point at which the demonization of the

Islam becomes the institutional philosophy of the newly formed Spanish Christian state.[167]

The Making of the Modern States of Spain and Morocco

Spain

With the eventual death of Isabella (1504) and Ferdinand (1516), the Spanish state entered a new and radically expansionist phase. Colon (Columbus) reached the Indies by 1492 and began Spain's western colonial expansionism in the interest of the acquisition of gold and in the name of the Church. While the Spanish pushed into the Americas, they simultaneously engaged in nearly a century of expansive incursions along the Moroccan coast. Carlos I, who had taken the throne from Ferdinand, was elected the Holy Roman Emperor (1519), and became known as Charles V, and now exercised sovereignty over the largest landholdings Spain ever controlled, including Burgundy, Flanders, the Netherlands, the West and East Indies, Naples, and Sicily. Meanwhile, Spain's Hernán Cortés had arrived in Mexico in search of gold and land (1519), to be followed by Pizarro and Almagro in Peru (1532–1533).[168]

While Spanish coffers grew full with the establishment of Buenos Aires in 1535 and the discoveries of silver in Columbia and Mexico (1545–1564), Spanish consolidation of its nearby geography was expanded. These holdings already included Melilla (a Spanish colony on the northern Moroccan coast) acquired in 1497 as well as Ceuta (also on the Moroccan northern coast and known as "Sebta" to its original inhabitants) in 1668.[169] Both these North African holdings were obtained from the Portuguese, either through outright seizure (Melilla) or treaty (Ceuta).[170] Both holdings were a great blow to the prestige of Morocco in the eyes of its rulers and population, especially in the wake of the events of 1492.

The seventeenth century saw Spain expand its conquests in South America, while it struggled to maintain its territories on European soil. These difficulties were exacerbated by the Plague's arrival on Spanish soil yet again, coupled with severe drought and famine (1581–1676). While Spain's greatly touted navy was humiliated by the English (1588), Spain's domestic regions reentered a period of military conflicts among themselves in a series of separatist wars (1631–1638). Meanwhile, Spain was forced to renounce its control of Flanders, bringing an end to both the Thirty Years War and Spanish hegemony in western Europe.

In 1668 the Treaty of Lisbon signaled the end of Spanish control of Portugal. By 1714, the Hapsburg line was replaced by the French Bourbons under Filipe V, which ushered in the royal family that continues, in name, to rule Spain to this day. The days of the Bourbons were not without strife. Soon Spain was drawn into the French Revolutionary War against Napoleon, which marked the beginning of a series of Spanish sufferings and humiliations including bread riots (1776), the French military occupation of Spain (1808), the replacement of its king with Napoleon's brother Jose I (1808), and the loss of the majority of its colonies. In 1813, the French withdrew, and the Bourbons regained the crown under Fernando VII, who reasserted Royal Absolutism. The Bourbons were to fall yet again at the death of Fernando (1833), beginning a period that led to the origins of the First Spanish Republic. This movement was quashed, however, as the army eventually rallied to the side of the crown, restoring the Bourbons once again under Alphonso XIII (1875–1876). Unfortunately for Alphonso, however, Spain had already acquired a taste for a republic, which resulted in a growing militancy in many areas of the country against the king. Meanwhile, Spain finally lost the last of its overseas territories in a confrontation with the U.S. Navy off Santiago de Cuba in 1898.[171] By this time, the power of the crown had waned, yet it was not extinguished before Spain secured a protectorate over western Morocco in 1912.

By 1931, Alphonso had ceded his power to the Second Republic, which attempted to institute extensive agrarian reform, in a country whose land had long been controlled by a small and powerful group of landowners.[172] These reforms were unsuccessful for a number of reasons, including their focus on the North to the exclusion of the South and the underestimation of the ability of the landowners to stop lower-class families from being placed on their land.[173] The results of the failures of the Second Republic led to widespread social unrest. While one response was the election of the Right in 1933, their reactionary policies drew voters back to the Left in the 1936 elections. Yet the Second Republic was to end all too soon, as a young Spanish officer stationed in Morocco named Francisco Franco (b.1892–d.1975) would use his base of power to raise an army and eventually crush the Republicans in a civil war (1936–1939). This lead to the beginnings of what would become an unbroken fascist control of Spain until 1975.

General Franco, or "Il Caudillo" (Supreme Leader) as he came to be known, allied himself with the Roman Catholic Church and portrayed his mission as one not dissimilar to the Crusades or the Reconquista: a tireless effort to rid Spain of "the infidel, the anti-Christ, the godless Republic of 'Jews, Masons and Communists,' the unholy alliance of the enemies of 'authentic, eternal Spain.' "[174] Franco's ascendancy was applauded by the Vatican, which eventually signed two concordats (1941 and 1953) with Franco, guaranteeing him the right to name all Spanish bishops, an act that has served to discredit the Church in the eyes of many Spanish people to this day. In turn, Franco reinstated many of the Church's privileges that had been stripped from it during the Republic, including that all levels of education in Spain be in keeping with Catholic teachings, the abolition of divorce, and the prohibition against any association with communism.[175] At the same time, Franco harnessed many of the symbols of Spanish Catholicism to fortify the legitimacy of his rule. Under the auspices of what came to be called "National Catholicism," Franco sought to fuse the aims of the state with the religious expression of the Church.[176]

For 36 years, Spain was shaped by the choices Franco made. Its neutrality, and in some cases outright collaboration with the Nazis, had far-reaching effects for Spain, isolating it for decades from the rest of Europe. Under Franco, however, there were areas of undeniable progress made, primarily in the building of an extensive middle class, something that did not exist prior to the Civil War. Many would attribute the longevity of Franco's regime to his profound ability to align himself simultaneously with both the Church and with principal Spanish business interests under the auspices of what some scholars call "corporatism."[177] Write Howard Wiarda and Margaret Mott:

> National identity in Franco's Spain reached back to medieval Catholicism. Those who were good Spaniards were those who fought for God, Spain and St. James. Liberals and Socialists, those vanquished during the (Civil) war, became the new breed of heretics.[178]

Thus, with an ideology that repressed dissent while tangibly rewarding cooperation, Franco was able to build a Spain that appeared to many to honor the country's past, while not making too many compromises with those labeled as "outsiders" (read democrats, socialists, communists, Muslims, Jews, and women who strove for their own liberation). In this regard, Franco heralded a return to the mentality of the Reconquista, celebrating an ethnic and cultural cleansing of the Iberian peninsula from all parties thought to be "non-Spanish."

Meanwhile, agriculture under Franco initially underwent a period of stagnation, as the new regime sought to undermine the agrarian reform instituted by the Republicans. Instead, well into the 1960s, Franco placed an emphasis on the small farmer, in keeping with his aim of preserving what he perceived as Spanish traditionalism.[179] This practice also led to the preservation of what were a limited number of crops normally cultivated by the small landholders.[180] The regime attempted to use those peasants who had been left landless in the wake of the abandoned reform movement as colonizers, but the numbers of those who benefited were minimal. At the same time, Franco mandated that prices be guaranteed for established crops such as grapes, olives, and wheat, a practice that stagnated experimentation in new areas of farming.[181] By the mid-1960s, however, the regime mandated a turn toward industrialization, which drew many people from the countryside to the cities. This phenomenon coincided with the disappearance of many small farms, which were replaced by much larger operations. These developments aside, Spain long remained among those European countries that produced a lower output of agricultural products than many of its neighbors. In recent years, however, this trend has been altered by the implementation of high-intensity greenhouse farming in the south, which now provides northern Europe with many "out of season" fruits and vegetables.

Franco's death in 1975 was the death knell for a system that had already been rejected by the majority of Spaniards, most of whom were eager to become active members of a greater and more modern Europe once again. Franco's ostensible control of the Bourbon monarch Juan Carlos I proved to be less than ironclad, as at Franco's death the king ushered in a transition government to prepare the country for eventual open and free elections, under the banner of what would become a democratic constitutional monarchy. In 1978, Spain ratified a new constitution and embarked on realizing a new system of government that would give an unprecedented amount of power to its domestic autonomous regions. By 1982, the PSOE or Spanish Socialists were elected under the leadership of Felipe Gonzáles. That same year, Spain joined NATO, and in 1986 it went on to join the EEC (now the EU). While the PSOE retained power in the 1993 elections, multiple allegations of corruption hobbled its effectiveness within its own borders. By 1996, the Conservatives (the Partido Popular or PP) won the general election by means of a coalition government under the leadership of José Maria Aznar. Aznar and the PP stood for reelection in March 2000 and won again.

Morocco

It was during the period between the ninth and thirteenth centuries that the entire Maghreb was unified politically.[182] While this period of geographic unity was not to last (the Maghreb ultimately being divided among what were to become the modern states of Tunisia, Algeria, and Morocco), it was instrumental in setting the groundwork for what was to become of Morocco in a post-1492 world. It was during this period that, contrary to many colonial accounts, the inhabitants of the western Maghreb were more than able to feed themselves without European help. Tenth-century accounts speak of an abundance of horticultural and mineral wealth. Writes Abdallah Laroui:

> The Andalucian al-Bakri (d. 1091/487) gives an account of the economic [and agricultural] situation. He describes at length the country around Fez and Oujda in northern Morocco, a thriving region served by several seaports: Ceuta, Tangiers, Badis, Mellila and Nakkur: Orchards were abundant— pears, pomegranates, and also vineyards; from Tetuán to Fez there were rich meadows, and in the vicinity of Basra cotton was grown; the region was also famous for its horse breeding. He also describes the southern region, taking in both slopes of the Anti-Atlas, its main cities being Masa, Igli, and the city (?) of Dra'; this was the region of the villages and the orchards; to the north of the Anti-Atlas the oil-bearing Argan tree predominated and sugar cane was grown. To the west of the High Atlas, in the vicinity of Aghmat and Nifis, there was an abundance of fruit trees, especially apple trees, while in the east the Sijilmasa region was rich in minerals as well as date palms.[183]

Yet it was following the expulsions of the Muslims from al-Andalous by the Spaniards in 1492 that set the stage for a variety of Maghrebi setbacks. While those who returned brought with them skills that were useful in augmenting Morocco's agriculture and industry, and founded key towns such as Tetuán, their return was an embarrassment. In 1496, this embarrassment was augmented when the Moroccan city Melilla was occupied by Spain. By 1515, Morocco was to undergo another occupation, this time by the Portuguese, who took for themselves a large swath of Morocco's Atlantic coast.[184]

During the ensuing years the so-called Maraboutic Crisis is said to have emerged in Morocco, pitting the power of the sultans, (aligned with what was then considered to be mainstream Sunni Islam) against the "walis" or "salihs," individuals who were viewed as local saints and who emerged from a syncretistic form of Sufism combined with indigenous religious traditions that embraced pre-Islamic ritual practice popular among Morocco's largely rural Imazighn (and sometimes Arab) population.[185] It should be noted, however, that this is an increasingly contested view of the events.[186] Walis or salihs, often referred to in Western literature as "marabouts" were integral players in a fascinating amalgamation of religion and politics that was emerging in Morocco.[187] Attributed as the source of revelation and miracles, the salihs wielded tremendous power among the agrarian class. A place of organizing power for the Sufis as well as the salihs were through institutions called "zawiyas," which acted as centers for religious education and gathering, and at times places that were a focal point of defensive warfare.[188] The countryside (and many urban centers) of Morocco ultimately became dotted with zawiyas, which acted as a decentralizing force against both Spanish incursion and the efforts of some Sultans to consolidate their power. People were drawn to the salihs due in part to the belief that they were God's intermediaries, who could convey "baraka," the grace or blessings of God, by their touch. Others were drawn by the fact that the salihs spoke in Tamazight, or Tachelhite (among other indigenous languages) rather than Arabic.[189] In a country already decentralized by competing tribal affiliations, the salihs and the zawiyas only served to frustrate urban efforts at uniting the rural peasantry under its own brand of urban-based Sunni Islamicization and political control. Meanwhile, while the salihs claimed a special status in relation to Allah, the Arab and later Imazighn Sultans claimed their own special status: as direct descendents of the Prophet Muhammad. Ultimately, so many different families of different tribes, ethnicities, and religious traditions claimed such a sharif (collectively called "shurafa") status, both by sultans (who were also at times known to refer to themselves as Khalifas or caliphs, depending on the dynasty) and salihs, the claim was made almost meaningless in the eyes of many Moroccans.

By the end of the sixteenth century, Morocco had seen a fascinating turn of political events. While the Turks and the Spaniards had competed for land and influence in the central and western Maghreb, the system of zawiyas supported a group of leaders who sought to construct a new unity across the Morocco in the wake of foreign incursions. Still, Morocco was composed of a variety of distinct regions, none of which held complete control of the country. These regions, two of which were governed by the northern coastal cities of Tetuán and Salé, shared in the Mediterranean economic system.[190] Ultimately, the sixteenth century brought with it the consolidation of many formerly independent regions under the Sa'di Dynasty, including Haha, Sus, Gazula, Haskura, and the Dar'a.

By the Seventeenth century, Morocco provided a significant amount of wheat to Europe via Portugal, along with sugar, wool, cloth, copper, horses, and salt.[191] Moroccans also traded "horses, wax, indigo, gum arabic and gold dust" with the Portuguese.[192] At the same time, Morocco was a trading hub for European goods headed to Sub-Saharan Africa, as well as an important trader of domestically produced cloth.[193] Politically, Morocco was composed of regions in possession of their own identities, principalities, and particular areas ruled by zawiyas, and shaykhdoms, such as the one found in Marrakech.[194] This autonomy was about to be changed, however, by a family from the southern town of Tafilalt, the Alawi, who were to alter the landscape of the entire country. After an initial attempt by Mohammed B. Ali from 1635 to 1659, the Alawi finally achieved their goal to fully unite Morocco under the armies of Ali's brother al-Rachid, between 1666 and 1671. The Alawi had no salih base of influence, but rather claimed sharifian powers and viewed themselves as the new sultans (and eventually caliphs) of Morocco.[195]

The Alawi Dynasty did not ascend without difficulties. Between 1727 and 1757 the Sultan 'Abd Allah was deposed five times.[196] 'Abd Allah's son, Mohammed III, learned from his father's misfortunes and reconstructed the role of the sultan more along religious lines, offering himself as a religious leader. Mohammed III reconfigured the tax base of Morocco, moving from regionally applied nonuniform systems to one with national aspirations. Mohammed III recruited for his armies strategically among the tribes, exempting some regions from taxation to quell their dissent. Mohammed promoted trade with Europe, built ports with foreign help, and could be seen as the legitimate architect, for good and ill, of the modern Morocco.[197]

The nineteenth century saw Morocco's economic fortunes decline as it became too heavily dependent upon the variations of foreign trade. Many, including Laroui, point to this situation as laying the groundwork for the eventual foreign occupation of Morocco.[198] By 1844, Algeria had been occupied by the French, and a Franco-Moroccan war resulted in the Treaty of Tangier, an agreement organized through English intervention.[199] Perceiving Moroccan weakness, Spain invaded Morocco in 1859, only to be pushed out by English pressure two years later. Fortunes changed, however, when in 1861 Morocco deepened its dependence on foreign trade by expanding its sale of wheat to include Spain along with its existing contracts with England. Extraterritorial privileges for foreigners were greatly expanded by the Franco-Moroccan Treaty of 1863, extending trade, exempting many foreign powers from Moroccan domestic laws, and laying the groundwork for Morocco's eventual partition by the European powers.[200] These decisions by the sultans were highly resented by the population, who did not economically benefit from the choices being made. The results included outright military revolts in Tetuán.[201]

Morocco's increasing dependence on foreign trade to fill its coffers was met by smaller and smaller returns. The overall effect was to weaken Morocco's ability to feed itself, undermine its economy, diminish its army, and lower the confidence of the population in its government. Morocco was now highly vulnerable, and its status was an open secret. By 1912, Morocco was placed under international control by the Treaty of Madrid, which established an official Spanish Protectorate in northern Morocco with its capital in Tetuán.[202] With the Treaty of Fez, Tangiers became an international zone, governed by the British, the Spanish, and the French.[203] It was at this point that the already significant international presence began to grow in Morocco. Also in 1912, the French began their own military drive to control the interior of Morocco, in some cases using Moroccans as recruits.[204] In the same year, new maps were being drawn that laid the groundwork for Spain's claim to what would come to be called the Western Sahara.[205] By 1914 the French controlled lands as far south as Taza, and the end of the bulk of Moroccan resistance was in sight.[206] In 1919, the French established a central government in Rabat to oversee Morocco. By 1931, there were over 172,000 foreigners (mostly Spanish and French) living in Morocco.[207] Writes Abdallah Laroui, "Throughout [what was to become] the colonial period, [Morocco] was almost always under martial law . . . The only public gathering place left was the mosque."[208]

While the French were conquering Morocco militarily, they were manipulating Moroccan law to claim ownership over vast tracts of land, which they in turn sold to industrial European interests.[209] The aim of the French was to develop large-scale farming tailored for European export.[210] At the same time a system of land registration was introduced that allowed French colonials to gain control over land formerly owned by Moroccans.[211] It was also during this period that European banks, under the auspices of taking over the sultan's European-induced debt, took full control of the Moroccan economic system. Yet despite all these tremendous blows to Moroccan pride, there was resilience among the population. The progression of popular confrontation with colonialism that emerged in the 1920s in Morocco included military resistance, religious reform, and political activism.[212] This military resistance culminated in the Riffian wars, carried out against Spanish troops from 1921–1926.[213]

While Morocco officially supported France in World War I, the average Moroccan was preoccupied with the challenges of daily living. A migration began toward the cities, where more work was to be found. Meanwhile many, but not all, religious leaders saw their job as having more to do with religious reform than battling colonialism.[214] While anticolonial activity was generally limited to urban areas, much of the countryside of Morocco remained to be pacified by the Europeans. By the end of World War II, the Protectorate was consolidated. It was at this time that the French even began to speak of the possibility of converting the Imazighn to Christianity.[215]

The 1920s and 1930s saw the development of what C. R. Pennell refers to as "capitalist agriculture," where dams were built and water was diverted from tribal lands to the colonialists.[216] Such indignities only augmented a growing sense of Moroccan nationalism, which was fed by the graduates of the Salafi "Free Schools," whose graduates returned to Morocco from Europe with advanced degrees (a number in agronomics) and a taste for confrontation.[217]

By 1934, the nationalists had gathered around the Alaw'i Sidi Mohammed, later to be placed on the throne as King Mohammed V. When he visited Fez in May 1934, he was greeted by large crowds shouting "Long Live the King!" and "Long Live Freedom!" despite the fact that the concept of a king was new to modern Morocco, which had been accustomed to be led by sultans.[218] Sidi Mohammed was a rallying point against colonialism, and would play a key role in securing Moroccan independence. The late 1930s and early 1940s saw the beginning of agrarian reform in Morocco. Previously Moroccan farms withered for lack of water while the colonials produced more grapes than they could use, but by 1938 the French began to sponsor irrigation projects for Moroccan farmers as well as low-interest loans.[219]

According to Pennell, "by September 1939, the colonial state in Morocco had laid the foundation of the national state which would succeed it."[220] By the end of World War II and the subsequent eleven years of colonialist violence, the Moroccans had "molded a nationalist movement in which a conservative monarchy was allied to radical supporters of democracy and revolutionary nationalism."[221] In the Spanish zone, tremendous hunger had brought great misery to the general populus, making it easy for the Spanish to recruit soldiers among those who wished to escape their bleak circumstances.[222] Still, the Moroccans remained steadfast on many fronts, something reflected in Sidi Mohammed's refusal to hand over Morocco's Jews to Vichy France.[223]

At the end of World War II, Franco's Spain was shunned by the former allied nations. Franco responded by seeking friends in the Arab world to compensate for his country's losses. At the same time, however, he treated Moroccans under Spanish rule in Morocco with unparalleled brutality. Meanwhile De Gaulle was trying to curry favor with the Moroccans by inviting Sidi Mohammed to attend France's victory parades, which naturally included a great number of Moroccan conscripts. By 1951, Sidi Mohammed was pressing openly for Moroccan independence, a plea that did not impress the French or the Spanish. Sidi Mohammed was thus becoming a liability to the French, who ultimately decided to deport him to Madagascar in August 1953. The results were entirely the opposite of French expectations. Rather than recognize the French-chosen successor, Moulay Mohammed Ben Arafa, Sidi Mohammed rose to a saintly level in the eyes of the average Moroccan, some women claiming to have seen his face in the moon.[224] Meanwhile, some of the most notable people who had fought against European colonialism in Morocco put together a plan to return Sidi Mohammed to Morocco. At the same time widespread acts of sabotage were being performed in many Moroccan cities, including the stabbing of the new European-chosen sultan on his way to Friday prayers in Rabat. By 1955, Moroccan, Egyptian, and Algerian militants were training the beginnings of a Moroccan liberation army in northern Morocco's northern city of Nador. The protracted violence and threat of still more violence wore down the French. On October 29, 1955 the French returned Sidi Mohammed to Nice, and on November 16, he returned triumphantly to Morocco. The French had thus unwittingly fortified the position of the sultan, and gave new life and longevity to the Moroccan monarchy. By March 2, 1956, Morocco achieved independence from France and in April of that same year, the Spanish gave in as well.[225]

With the coming of independence, Sidi Mohammed, now called King Mohammed V, had to face an entirely new set of domestic challenges.[226] Those who had united to fight the French and Spanish were not necessarily inclined to remain allied to an independent Morocco. Further, as Pennell has noted, "there was no agreement on the specific shape that Moroccan society should take after independence."[227] Having originally promised a constitutional, democratic monarchy, King Mohammed V moved ahead to consolidate his power in the face of multiple opposition parties. The power vacuum thus left by the departure of the Europeans was filled by Mohammed V alone, without a legislature, and without a Moroccan bureaucracy.[228] Mohammed V declared Sûrete Nationale, an act that would eventually lead to the formation of an extensive domestic police force charged with maintaining the power of the throne. Mohammed's eldest son, Hassan, was appointed the head of the army. The political party Istiqlal, one of many prior to independence, became an arm of the throne, and thus gained considerable influence. By 1958 Istiqlal dominated the entire Moroccan government.[229]

1958 also saw an impressive military rebellion in the Rif against the new government, which was eventually crushed by Hassan's forces. It was also during this period that the king attempted to impose land reform. However, Mohammed V's version of land reform in many ways retained the inequality of the European systems, which placed small farmers and their parcels under the umbrellas of larger units controlled by the government, who owned the tractors and the means of distribution. Morocco's continued reliance on Europe was also reflected in its economic commitments to its former occupiers, something that greatly hindered Moroccan development.

The following years saw the throne consolidate its power in both rural and urban settings, while it successfully manipulated independent political parties to increase the King's power.[230] In 1961, Mohammed V died, and was immediately succeeded by his son, who became King Hassan II. The first Moroccan Constitution, declared in 1962, left the majority of power in the hands of the throne. During his rule, Hassan II faced multiple opponents, and more than a few attempts on his life. His response to such opposition was swift and sure, killing many and leaving more to languish indefinitely in Saharan prisons. It is fascinating to note that among those Hassan imprisoned were people aligned with the socialist and communist parties. Imprisoning these individuals gave Morocco's new king something in common with Spain's leader to the north, as Franco did not flinch from arresting, and in some cases "disappearing" those thought to be socialists or communists. At the same time Hassan II was brilliant in his ability to link his position simultaneously to the state as well as taking on the title "Commander of the Faithful," placing the throne at the head of an ostensibly unifying form of syncretistic Sunni Islam. Hassan II thus became not only a statesman, but also a religious leader of "traditional" Islam, while simultaneously acting as a mediator of the same baraka once only offered by the salihs.[231]

Despite his many powers, Hassan II presided over a country whose economy still remained tied to European interests. One such example can be found in the government's determination to focus on highly irrigated for export crops at the expense of domestic agriculture.[232] Such tactics as these were employed by Hassan on many fronts, and negated his popularity with the Moroccan populace. However, in 1975, he came upon the brilliant idea of the Green March, where he implored 350,000 Moroccans to march upon the Western Sahara and force the Spanish to abdicate their long-held territorial position. With the death of Franco on November 25 of that year, Spanish control in this southern region came to an end. Moroccans were jubilant; in the eyes of many Moroccans Hassan II had restored much of the prestige he had lost over the years. While the issue of the Western Sahara is still in contention to this day (it is claimed by indigenous forces represented by the Polisario, as well as Algeria), it is controlled by the Moroccan military.

In the ensuing years Hassan ostensibly diminished the scope of his powers by opening a Parliament, although he retained complete veto rights for the majority of his rule. He also positioned himself as a statesman, offering Morocco as a mediator between the Palestinians and the Israelis, looking to the West for support in exchange. Hassan II, like the leaders of many developing countries, had his own struggles with the IMF, but he was not deterred from continuing in his quest for Morocco to become a player in the Northern political system. In this spirit, he applied Morocco for membership in the EC in 1987.[233]

Hassan's tight grip on power in Morocco led him to be seen by Spain and others as a "reliable" Muslim ally, who was ostensibly able both to prevent the growth of fundamentalist Islam on Moroccan soil, and to remain in friendly dialogue with the North.[234] By the time of his death in 1999, Hassan II could count among his accomplishments a relative opening up of Moroccan society, with some modest progress made toward more pluralism and free speech. While Hassan's attempt to form a trans-Maghreb union was ultimately a failure, it may still bear fruit in the future political development of the region. Finally, Hassan II was able to give his throne to his first son, who became King Mohammed VI in July 1999. The young king is both European-educated (with a Ph.D. in International Law), and most likely half Imazighn.[235] It will be up to him to either chart a new course for Morocco under a more pluralistic system, or risk the end of the Alawite Dynasty.

Three Areas of Conflict, Three Opportunities for Cooperation

The challenges that face modern Spanish–Moroccan relations are clearly rooted in their intertwined Ecological Histories. Conflicts over land, natural resources, and immigration are currently among these two nations' greatest points of contention. Yet from the point of view of Ecological Realism and Sustainable Diplomacy, these conflicts also offer themselves as opportunities to strengthen ties across the Strait of Gibraltar, while bridging the chasms of the North–South split, Muslim–Christian conflict, and the gross disparities between the wealth of the EU and Maghrebi poverty. Thus it is these three areas, *land*, *natural resources*, and *immigration*, which will occupy the work of our next three chapters.

CHAPTER 3

THE CONFLICT OVER LAND: THE FIRST HUMAN, LAND USE AND THE TWO CITIES

This is the work of Sustainable Diplomacy: to identify the potential origin and elementary vocabulary of a common language between two peoples who have been for too long separated by distrust, human language, differing religious interpretations, and an anthropocentric history. Sustainable Diplomacy must then turn to the task of using this new language in actual negotiations and other forms of community building.

The proponents of Sustainable Diplomacy are obliged to mine two sources of language for their work. The first common text that both sides must read together is the text of the land itself and all its inhabitants be they human or nonhuman. This is a text that must be read again and again, as modern developments such as the Ecological Footprint analysis present themselves, and as both sides learn to understand themselves, their neighbors, and their ecosystem on a deeper level. The practitioners of Sustainable Diplomacy must then turn to the sacred texts of one another's traditions beginning with the Creation accounts, and read them in the new light that reading the land has provided. The combination of these two readings will suggest common principles. From these principles the framework and guidelines for a practical approach to conducting Sustainable Diplomacy will emerge. There are numerous similarities that each population can see and in many cases potentially agree upon. Thus the combination of the common ecological texts and the sacred written texts must be seen for what they are: invitations to build vital connections, places of illumination, subjects for continuing reinterpretation, and new sources of approaches to cooperation.

Reading the Land: Interviews with Antonio Dueñas Olmo and Rabha Ahmed

The individuals presented in the following interviews possess knowledge of the subject under discussion—the land and its traditions. Each individual presents glimpses of times that have long passed and yet somehow impact the present. Antonio Dueñas Olmo, a Spaniard who was born in northern Andalucia, is a 43-year-old secondary school teacher of art history in Seville. In sharp contrast, Rabha Ahmed was born in the Sahara before the French occupation of Morocco in 1912. At the age of 90, she now lives in Khenifra and retains memories of different and distant times in the story of her country.

Antonio Dueñas Olmo

When you hear the word "land," or particularly "the land of Spain," what images come to mind for you?

Every time I think about land, I think about plowed fields, the fields with the furrows that have been plowed to be sown, and tones of red, generally with little vegetation. My land, I would imagine, is similar to the fields of Andalucia. You know that there is an Andalucia that is very green and fertile, and an irrigated Andalucia, but I always have that first image in my mind. Lots of soil, little vegetation, a man or a woman with hard wrinkles in their faces from the sun and work, directly associated to the hard land and difficult work.

What do your religions and/or cultural traditions teach you about the origin of Creation?

Tradition has taught all of us, and I say all of us because we have all been born in to Catholic families, what the Bible says—that we are all born from Adam and Eve. It is clear that we have overcome this legend or mythology of the Bible. But this is what tradition says, that, in some way, we directly come from God.

Where did you learn about these traditions?

Mainly at home, at church, and at school. You can imagine how it was in a Franco school. When I went to high school, I had to pass a religion exam and one of the questions was who were our first parents. Of course, you had to answer Adam and Eve, in a very Catholic, Apostolic, and Roman manner, which was supported by the powers of Franco. You weren't allowed to say that man comes from the monkey by means of evolution; that was forbidden. Even nowadays, in some Catholic schools, like those of the Opus Dei, the teaching of evolution is forbidden. Darwinism is totally rejected because they consider religion above everything else.

Did you or will you teach these stories to your pupils?

No, not at all. I talk about this because I like to talk about it, and my religious formation has been useful to establish comparisons, in relation to mythology. Each and every civilization has its mythology. I talk about the Christian mythology and when we talk about the saints, well, rather about the characters in the Bible. I see them as mythological elements, although some of them have really existed.

Is there any specific mythology in the place you come from?

Well, it may be, but these particular mythologies have been overridden by Catholicism to the point that today they cannot be followed. But, for instance, I'll give you a specific example. In my village there is a religious vocation for Our Lady of the Moon, or the Virgin of the Moon. Logically, the moon is a heavenly body that is closely related to fertility. Nowadays, in this worship of this virgin, you can actually see certain traces of old festivities related to stars, that is, to the forces of nature. And this is not just something in my village. There are virgins all over the region such as Our Lady of the Star, Our Lady of the Paths, Our Lady of the Holy Stone, and they are all elements that make reference to nature. They are the forces of nature that have been covered by the Catholic religion, and that great Mother Earth has been transformed to become the Virgin.

Are there songs or poems?

Yes, there are some songs, legends, and even Iberian dances, in a nearby village called San Benito, where they dance with swords surrounding a saint, San Benito. Some anthropologists say that these dances are old Iberian fights.

Then, the Christian traditions adopt older traditions?

Yes, of course, but they transformed them. But it is clear that people don't easily give up traditions. So Catholicism transformed them, but inside they are the same. I think that nowadays celebrations are similar to the old pagan ones where food, drink, and dance were really important, as they still are.

Do you think that people establish these differences between their traditions or the (pagan) traditions? Do they differentiate between them?

I think that some people may. Religion has always had a lot of power over the peoples' minds, but I consider that sometimes this distinction is made, mainly those people related directly to the land, especially the farmer. Their time is not based upon a calendar of festivities, but upon the fundamental laws of nature, because they have to water, because the sun is shining . . . My grandfather was a person that when time changed (daylight savings time), he just ignored it. He always used the sun as his clock. He always woke up at dawn before the sun came up; it didn't matter the time of day it was, then I think that people connected to the land, to farming, to agriculture are controlled by the forces of nature, by the sun, because they have to water, by the rain, those elements.

Antonio Dueñas Olmo, 43, teacher, Seville[1]

Rabha Ahmed

You lived in the countryside and you came to the city. Do you prefer life in the countryside or in the city?

Every time has its different places, different times . . . In the countryside, there used to be sheep and agriculture and everything . . . Here too has its rules. We were better off in the countryside. Comfort and well being here is different from in the countryside.

What do you prefer?

In the countryside, there were both hardships and happiness. Here there is both laziness and work . . . If you have nothing [in the countryside] you bring sheep or something to the souk [and sell them]. Here if you have nothing you won't be able to eat.

Isn't this true? But this is when life was good. But now, there is no rain . . . .Once, we went out with my brother-in-law. As I'm short, we were walking in a field of wheat and [I] was hardly seen. So [my brother-in-law] was wondering if I was standing or sitting! There were all these rains and . . . there is nothing now.

There are no rains

This is because of the sons of Adam, and jealousy. Now people mark their lands with stones [a primitive form of enclosure] so that the animals won't graze. In the past, you sent your sheep and cattle to graze wherever they liked. Man became like [this] so God made it hard for him . . . I'm sorry for being so talkative. Is what I said true or not? This is all because of man. Women used to give birth to two children or one . . . Me for example, I gave birth to many children but I never went to see a doctor or went to a hospital. I did it with the help of God. Now women go always to see a doctor; they don't seek the help of God. The situation has changed.

When you hear the word "land" or the phrase "the land of Morocco," what image comes to your mind?

If the land is mine I can sell it if I wish; otherwise I'll keep it and profit [from] its products. If it's not mine I don't care a straw. I can't interfere with what is not mine . . . I can't interfere with that and with Morocco as a whole. That the Moroccans win or lose is none of my business. I care only for what is mine. What is far from me I keep far from it. When something is close to you you stick to it . . . Isn't this true?

What does your religion, Islam, teach you about the origin of the land? When was it created?

God created everything. God created the land, stones, trees, and everything. Nowadays, people would uproot a tree and plant it somewhere else. In the past, trees [and other things] were left as they were originally created by God. Isn't this true?
Rabha Ahmed, 90, widow of a farmer, Khenifra[2]

Reading the Sacred Texts: The Creation of the First Human

The common stories, textual accounts, and popular myths of Creation arguably act as common ties that help to connect and form the identity of a religious culture, regardless of individual beliefs or disbeliefs. For Muslims and Christians, contemporary human ties to the land have long been presented as a reflection of the ties in which the first human's creation forcefully bound animate life to seemingly inanimate soil, clay, or dust. Thus the emergence of the human may be understood as having taken place within the folds of Creation itself, regardless of the particular theological anthropologies that ensue.

For the purposes of comparison, one must heavily weigh what has become the popular account of these Creation stories, rather than favoring the work of scholars. While it is scholarship that can invite us to go beyond the difference in view between two or more peoples, it is the stories of the people themselves with which one must always begin.[3] As we have seen, for Antonio Dueñas Olmo, the Biblical accounts of Creation have become mythic, as they have for many among the younger generations of Spaniards. Franco's perverse blending of Church teachings and modern fascism forever changed the theological landscape of modern Spain. Many of those who lived through the Franco years and those who have followed have lost a great deal of respect for the Church and its teachings — a Church that openly supported and gave its blessings to a man who allied himself with Hitler.

From the Moroccan perspective as well, the notion of blending religious teachings with state power is not a novel concept. The Alawi Dynasty, like those before it, has used the power of Islam to fortify its claim to the throne while it simultaneously pacified the national populus. While many if not most Moroccans would be loath to call themselves "nonbelievers," many are skeptical of the faith of those who lead them. While Hassan II took on the title "The Commander of the Faithful" and tied the newly instituted title of King to that of religious leader or Khalifa, many initially remembered him more for his reputation as a young playboy rather than a religious contemplative. As M. E. Combs-Schilling has noted, however, it was through the timeless rituals of Moroccan Islam that the modern sovereigns of Morocco were not only established, but in fact continued to retain power and prosper.[4] Hassan II thus emerged from his previously perceived persona to that of a great religious and political leader, a tradition firmly established and passed on to his son, Mohammed VI.

However religion is used at the state level, its real potency remains dependent upon those at the most local levels: to those who still pray, to those who make pilgrimages, and even to those who would not call themselves believers of any religious tradition. Thus the Creation story becomes a common language with different levels of meaning. The story of Adam, or the first human, is a touchstone, even if it is often only perceived as a common myth. For while Rabha Ahmed might view the story of Adam as the literal truth regarding her origins (as would many Moroccans), Antonio Olmo still concedes that at the root of many of our myths lie individuals "some of [whom] have really existed." Thus the story of Adam, even as a metaphor, offers to those who seek Sustainable Diplomacy a common language with which to begin a conversation. It is a conversation that acknowledges both the historical and contemporary power of belief and a grounded respect for the unique cultures and peoples it intends to engage.

The First Human and Creation in the Muslim and Christian Scriptural Canons

From the Bible

> And God said, "Let the waters under the sky be gathered together into one place, and let the dry land appear." And it was so. God called the dry land

Earth, and the waters that were gathered together [God] called Seas. And God saw that it was good. Then God said, "Let the earth put forth vegetation: plants yielding seed, and fruit trees of every kind on earth that bear fruit with the seed in it." And it was so. The earth brought forth vegetation: plants yielding seed of every kind, and trees of every kind bearing fruit with the seed in it. And God saw that it was good. And there was evening and there was morning, the third day.

—Genesis 1:9–13[5]

Then God said, "Let us make humankind in our image, according to our likeness; and let them have dominion over the fish of the sea, and over the birds of the air, and over the cattle, and over all the earth, and over every creeping thing that creeps upon the earth." So God created humankind in [God's] image, in the image of God [God] created them; male and female [God] created them. God blessed them, and God said to them, "Be fruitful and multiply, and fill the earth and subdue it; and have dominion over the birds of the air and over every living thing that moves upon the earth."

—Genesis 1:26–28

Then . . . God formed [the human] from the dust of the ground, and breathed into [the human's] nostrils the breath of life; and the [human] became a living being.

—Genesis 2:7

So out of the ground . . . God formed every animal of the field and every bird of the air, and brought them to the [human] to see what [the human] would call them; and whatever the [human] called every living creature, that was its name.

—Genesis 2:19–20

From the Qur'an

Surely your Lord is God who created the heavens and the earth in six spans of time, then assumed all power. [God] covers up the day with night which comes chasing it fast; and the sun and moon and the stars are subject to [God's] command. It is [God's] to create and enjoin. Blessed be God, the Lord of all the worlds. Pray to your Lord in humility and unseen. God does not love the iniquitous. And do not corrupt the land after it has been reformed; and pray to [God] in awe and expectation. The blessing of God is at hand for those who do good. Indeed it is [God] who sends the winds as harbingers of auspicious news announcing [God's] beneficence, bringing heavy clouds which We drive towards a region lying dead, and send down rain, and raise all kinds of fruits. So shall We raise the dead that you may think and reflect. The soil that is good produces (rich) crops by the will of its Lord, and that which is bad yields only what is poor. So do We explain Our signs in different ways to people who give thanks.

—Al-A'raf 7:54–58[6]

Remember, when your Lord said to the angels: "I have to place a trustee on the earth," they said: "Will you place one there who would create disorder and shed blood, while we intone Your litanies and sanctify Your name?" And God said: "I know what you do not know." Then [God] gave Adam knowledge of the nature and reality of all things and every thing, and set them before the angels and said: Tell Me the names of these if you are truthful.' And they said: "Glory to you (O Lord), knowledge we have none except what You have given us, for You are all-knowing and all-wise" Then [God] said to Adam: "Convey to them their names." And when he had told them, God said: "Did I not tell you that I know what you disclose and what you hide? Remember, when We asked the angels to bow in homage to Adam, they all bowed but Iblis, who disdained and turned insolent, and so became a disbeliever."

—Al-Baqarah 2:30–34

For God the likeness of Jesus is that of Adam whom God fashioned out of dust and said "Be" and he was.

—Al'Imran 3:59

And when I have fashioned him and breathed into him of My spirit, bow before him in homage.

—Al-Hijr 15:29

At first glance, the Christian and Muslim stories of origin regarding the first human appear to jibe with one another: in both accounts a creature literally made from the Earth is given the breath of life by the one true living God, and bestowed with the rights and responsibilities of one who is created in the image of its Author. While many aspects of the Muslim account of human creation are congruent with the Hebrew Bible account, in some respects the Qur'an moves beyond Biblical claims and endows the first human with a quality of knowledge, a role, and abilities that are not evident in the Hebrew text.

Despite the clear differences, the common ethical claims that are shared by each account of the creation of the first human are striking and multifold. It is within these common ethical claims that one can find a basis for conversation between the Christian and Muslim traditions, and it is from these claims that the practitioners of Sustainable Diplomacy can begin to fashion a common Earth-centered language with which to forge agreements and cooperation across the boundaries of differing religious cultures.

Common Ethical Principles : The Role of the Human and Creation in the Qur'anic and Biblical Creation Accounts

1. The human, the Earth, and all its creatures are in a common, interrelated ontological relationship

In both the Biblical and Qur'anic accounts, the human is physically embedded within the folds of creation by a God who fashioned the human from the very substance of the Earth itself. What was inanimate became human and animate through the artistry of a creator God. Eventually, the body of the animate human will return to the Earth upon its earthly death.[7] Thus, despite having a particular and distinct role, the first human emerges from a material common to all living things, and is fashioned by the same creator God who fashions all of the human's fellow Earth creatures. The human is therefore intimately tied and dependent upon the greater Earth community in both the Christian and Islamic Scriptures.

2. *The human is in an intimate, evolving and dialogical relationship with the Creator God. This relationship is shaped by human action and the human's ability to understand the meaning of responsibility to greater Creation*

In both the Biblical and Qur'anic accounts, God speaks directly to the first human. God's words describe the human's origin, the human's needs, and the tasks that the human must take on. These tasks include giving the other creatures their names and obeying God's requirement to refrain from eating of a particular tree. In both traditions, the human is asked, in one manner or another, to act as a representative of God on Earth. In the Qur'an, this responsibility is reflected in Adam's being made God's "trustee on earth" (Al-Baqarah 2:30). In the Bible, similar human responsibility is reflected in the human being made "in God's image," while being called to have "dominion" over the Earth and to "subdue it" (Genesis 1:26–28). In both cases, the human acts as a locus of divine hope and potential earthly peril, especially when the humans break covenant with God and eat from the tree that God had forbidden. All these activities take place through a human who has had a direct relationship with the Creator God who has spoken with the human. Such direct interaction with and knowledge of God is a privilege that is not apparently accorded any other member of Creation. It is through this initial human–Divine relationship that rules are made, patterns are established, and the qualities of the Divine are reflected in a new way upon Creation. At the same time, the earthly finiteness of the human is also described: it is a creature whose body becomes a temporary vessel for a spirit that will eventually survive the body's participation in the cycle of earthly life and death.

3. *The human is in an intimate and evolving dialogical relationship with human and nonhuman Creation*

In both the Biblical and Qur'anic accounts, Adam is given the privilege of bestowing the names of the other Earth creatures upon them. The Qur'an teaches that these names were given to Adam by God, while the Biblical account implies that it was Adam who created the names (Al-Baqarah 2:31; Genesis 2:19). The first human is also matched with a companion in a manner that differs in the two texts. While the Bible gives two accounts of the creation of the first woman, Eve enters the Qur'an without an explanation of her origin. In both cases, it is clear that the first humans were designed to live in a harmonious manner with the rest of God's Creation. When

Adam and Eve are forced to leave the garden from which they first emerged, both the Qur'an and the Bible charge them to work together despite any and all difficulties, while learning to live by their wits in a far larger and until now unknown Creation. While their earthly paradise is lost to them, Adam and Eve are by no means separated from God's greater Creation. In fact, one might argue that it is through the act of the ejection from the garden in both the Qur'an and the Bible that the first humans become even more deeply embedded in the folds of earthly Creation, toiling to feed themselves while they strive to fulfill their role in populating the Earth and acting as God's representatives upon the planet. It is within this Creation that they will encounter each other in a new manner and learn through trial and error the qualities and configurations of God's Creation.

4. The human is in possession of tremendous power vis-à-vis Creation, bestowed by God's very design

While the Qur'an teaches that the human is to be considered God's "trustee" or Khalifa (highest representative) upon the Earth, the human of the Biblical account is charged with "subduing" the Earth and having "dominion" over it (Al-Baqarah 2:30; Genesis 1:26–28). In both cases, the human, having been made either in God's image (Genesis) or in the Islamic tradition as God's own earthly reflection, is given extraordinary powers compared to those of the other earthly creatures.[8] Clearly, the human has the power of choice, and with choice comes the power to submit to God's will or to defy God, along with the power to live in harmony within the folds of Creation or act as an agent of its destruction. Thus it is within the power of the human to choose a path that could affect all other earthly creatures, and ultimately the biosphere as a whole. In this way, the human lives within Creation but is set apart from Creation's other members. In both the Qur'anic and Biblical accounts the human is in this respect a liminal figure, the locus of Divine hope, and the potential agent of earthly destruction—the pivot upon which turns a great deal of the earthly fate of God's Creation. The human's ultimate choices thus emerge from within the tension of human egoism and the Divine design that invites submission, cooperation, and respect for God and for Creation. These choices will therefore be made, in both the Muslim and Christian accounts, in a context in which the human is given the choice to reflect God's presence upon the Earth or forget and forsake all of God's blessings bestowed and revealed through God and God's Creation.

5. There is a language within and expressed by Divine Creation

In both the Christian and Muslim accounts, the biosphere in which Adam lives is itself a direct expression of God's will, purpose, and design. In this respect, one can view Creation itself as a form of Divine lingual expression. In both the Biblical and Qur'anic accounts, God is seen as speaking Creation into being (Al-I'mran 3:59; Genesis 1:3). In this way, one can see that Creation is itself among the most unimpeachable texts God has given to humanity. God's voice calls Adam into being in the Qur'an, just as God's utterances separate the light from the darkness in the Bible. Thus Creation itself becomes a living record and an earthly testament to the speech, designs, and desires of the Divine. Further, while it is accomplished in different ways, both the Qur'anic and Biblical accounts of Creation attest to a God who implores the human to learn and understand Divine intent through interacting with the text that is Creation. Thus, while the human is portrayed as enjoying particular favor from God and particular responsibilities vis-à-vis Creation, the human is also given strict instructions to mediate God's will upon a living and dynamic Creation that is precious in and of itself to God. Thus the value of God's speech is embodied in the importance of the responsibility placed upon the human to preserve what God has created. For Muslims and Christians, Creation is full of Divine signs or, if you will, full of Divine language. This language is expressed in seasons, in the growth of plants and trees, in rocks, in the sky, in an animal's behavior, in taste, touch and smell, and in other humans. Reading the text that is Creation is therefore among the most important tasks of the human from both an Islamic and Christian perspective. The text of Creation is always speaking to the human, and the human in turn is always, for good or ill, in conversation with Creation.

Points of Disagreement: The Role of the Human and Creation in the Qur'anic and Biblical Creation Accounts

While it is critical to keep in mind the commonalities offered by the Bible and the Qur'an regarding Creation and its first human inhabitants, it is equally important to keep these commonalities in tension with the disagreements between the two texts. These disagreements manifest themselves in three principal areas:

Naming

There is a clear discrepancy between the Qur'anic Adam and the Biblical Adam with regard to the act of naming the animals (Genesis 2:19; Al-Baqarah 2:33). In the case of the Biblical account, we find Adam invited by God to name the other Earth creatures from his own imagination. In stark contrast, the Qur'anic Adam, who has been given "knowledge of the nature and reality of all things" has clearly been given by God the names of his fellow creatures, which is information kept from even the angels. Some interpreters of the Qur'an have gone on to add that in fact, the names God gave to Adam were (or also included) the names of God.[9] While both the Biblical and Qur'anic Adams convey a certain degree of power in having the right to name their fellow creatures, the Qur'anic Adam, we are told, has been given far more knowledge (and implied power) than this. The knowledge of all reality is a level of knowledge the Biblical Adam will never achieve in his lifetime, and he thus stands clearly below the level of angels in his ability to affect Creation. In contrast, the Qur'anic Adam is clearly a marvel, who merits the bows of the angels themselves. The Qur'anic Adam, writes Ibn al-'Arabi, fails to heed the advice of God regarding the tree because he has the capacity to forget the knowledge he has been given.[10]

Sin

Doctrinally, the Adam of the Qur'an and the Adam of the Christian inter-pretation of the Bible occupy completely different roles vis-à-vis the humans who follow them. Foundational Christian doctrine teaches that the "sin of Adam" enacted by the eating of the fruit of the Tree of Knowledge of Good and Evil comprises a level of sin, which is passed along to all of humanity, only to be expunged by the life and death of Jesus of Nazareth. In stark contrast, the sin of eating the fruit in the Qur'an is Adam and Eve's alone, and occurs due to their "forgetting" what God had told them. In both cases, the couple is ejected from the Garden and told that their lives will now become appreciably harder (Genesis 3:22–24; Al-Baqarah 2:36). In many respects, Adam is, in both accounts, an "every-man" subject to the same flawed decisions all humans inevitably make with regard to their relationship to the Divine and to Creation. Both humans exhibit the ability to learn from experience and to discern right from wrong.

Role of the Human in Creation

Clearly, the Qur'anic Adam occupies a grander role at the onset of his Creation. The Qur'anic first human is made a "trustee" or "vice-regent," acting as a representative of God on Earth. By tradition, the Adam of Islam is also considered to be a prophet, one who will lead others to a belief in monotheism. The Biblical Adam is certainly not without his importance as well, as he is described as having been made in the "image of God" and instructed to "have dominion" over the other creatures of the Earth and "subdue" Creation (Genesis 1:26–28). Nonetheless, for all his ostensible importance, the Adam of the Bible is unable to take responsibility for his actions in the Garden when confronted by God, and he pathetically blames Eve for his transgression. In contrast, the Qur'anic first couple eat off the fruit together as equals and are subsequently ejected from the Garden to the same degree and quality of punishment.

The Ecological Footprints of Morocco and Spain: The Land

The human's responsibility to Creation is clearly pivotal in both the Biblical and Qur'anic accounts. Regretably, our most reliable means of measuring the human impact upon Creation leads to the conclusion that the human population, by both Muslim and Christian standards, has not fulfilled its earthly responsibilities. This conclusion is clearly reflected in the Ecological Footprints of Morocco and Spain. In order to demonstrate this fact, we first measure the footprint in terms of the amount of land that the Spanish and Moroccans utilize in order to meet their consumption needs. This analysis will stand as a touchstone for comparison, contrast, and potential cooperation between Morocco and Spain for this and the next three chapters. Using Wackernagel and Rees' most basic Ecological Footprint analysis with regard to the land, we must first determine the four categories of land and land use—Energy Land (or land appropriated by fossil energy use, or the land area required to sequester the CO_2 produced by the respective country); Consumed Land (land that is built upon or otherwise degraded); Currently Used Land (including gardens, crop land, pasture land, and managed forests); and Land of Limited Availability (including untouched forests and nonproductive areas such as deserts or ice caps).

Spain

In the Ecological Footprint analysis of Spain, a portrait emerges of a country that is among the top end of northern nations in terms of consumption and waste generation. With a land area of 49,994,000 ha, Spain has far exceeded its domestic capacity to produce and absorb the impact of its level of consumption. In terms of *Energy Land*, Spain currently emits 232,484, 500 metric tons of CO_2 into the atmosphere, an amount that would require 129,158,0567 ha of land made up of carbon sinks to absorb. The area of carbon sinks required to absorb this amount of CO_2, typically forested land, exceeds by over two-and-a-half times the size of the entire country. The land within the Spanish borders that is prepared to absorb CO_2 is actually only a modest 8,388,000 ha.[11] In order to compile the Ecological Footprint of Spain's Energy Land, it is necessary to break down Spain's emissions and ability to absorb them as a hectares per capita (ha/cap) statistic. Thus, the hectares per capita of CO_2 emissions among Spaniards, or Spain's Ecological Footprint is 3.2 ha/cap. In stark contrast the Ecological Footprint (or hectares per capita of land) available to absorb CO_2 output is a highly inadequate .21 ha/cap.[12]

The land in Spain that comes under the rubric of *Consumed Land* includes all land that is considered to be degraded, principally that land which is built upon—including both human settlements and roads. When broken down into a per capita figure, Spain divides its degraded land among its population at 497 square meters per person, making its ultimate Ecological Footprint for Consumed Land to be .0497 ha/cap.[13]

The third category of land use, *Currently Used Land*, includes cropland, managed forests, pastures, and gardens. Spain's arable cropland totals some 19,164,000 ha. Considering factors of importation and exportation of agricultural goods, including cereals, roots and tubers, pulses, vegetables and fruits, tobacco, sugar, managed forests and pastures, the average Spaniard's Ecological Footprint for Currently Used Land breaks down to .286 ha/cap for cropland, .34 ha/cap for forests, and .64 for pastures for a total footprint of 1.266 ha/person.[14]

The final category of land under the Ecological Footprint rubric is that of *Land of Limited Availability*, which includes untouched forests (which are seen as productive natural ecosystems for their CO_2 absorption capability) and nonproductive areas, such as deserts and ice caps. For the purposes of this study, taking into consideration the available data, I have defined "desert" as those lands that come under the headings of "arid" and "semiarid," while acknowledging the limitations of such a definition.[15] Totaling some 19,789,900 ha, Spain's desert area footprint totals .499 ha/cap. Spain's untouched forests have been greatly diminished over the last few years, and now total a paltry 13,556,652 ha, making a footprint of only .34 ha/cap. Finally, Spain's polar and alpine territory, the last component in the Land of Limited Availability, totals 998,880 ha, or .03 ha/cap when divided among the entire Spanish population. Added together, the total Land of Limited Availability for Spain totals 1.189 ha/Spaniard.[16]

Taking into consideration all four of the above categories, the student of Ecological Footprints is left with a very clear picture of Spain's consumption rates versus her ability to provide for her human population's needs within Spain's own borders. While Spain's actual area is 49,944,000 ha, the land required to provide all the goods and services Spain consumes is a decisively larger 178,320,000 ha. Thus, the difference between Spain's actual area and the area it requires for its level of human consumption is 128,376,000 ha, over two-and-a-half times the size of Spain. With these statistics, it is now possible to calculate the *Ecological Footprint of the average Spaniard*, which is 4.5. This statistic stands in stark contrast to what has been calculated as a Fair Earthshare if the world's resources were equally divided among its human population, a calculation that leaves each human worldwide with 2.1 ha/cap. Therefore, Spain exceeds the world "fair share" by 2.4 ha/Spaniard.[17]

Morocco

The Ecological Footprint analysis of Morocco reveals a country that is more typical of a Two-Thirds World country in terms of consumption and waste generation. With a land area of 44,630,000 ha, Morocco has only slightly exceeded its domestic capacity to produce and absorb the impact of its level of consumption. In terms of Energy Land, Morocco currently emits 27,879,400 metric tons of CO_2 into the atmosphere, an amount that would require 15,488,556 ha of land made up of carbon sinks to absorb. The area of carbon sinks required to absorb this amount of CO_2, typically made of forested land, exceeds by over three times the amount of actual land within Morocco's borders that is capable of performing this task. The land within the Moroccan borders that is prepared to absorb CO_2 is actually only 4,079,000 ha.[18] In order to compile the Ecological Footprint of Morocco 's Energy Land, it is necessary to break down Morocco's emissions and ability to absorb them as a hectares per capita statistic. Thus, the hectares per capita of CO_2 emissions among Moroccans, or Morocco's Ecological Footprint for Energy Land is .41 ha/person. In stark contrast the Ecological Footprint (or hectares per capita of land) available to absorb the Moroccan CO_2 output is an inadequate .21 ha/cap.[19]

The land of Morocco that comes under the heading of *Consumed Land* includes all land that is considered to be degraded, principally that land which is built upon—including both human settlements and roads. When broken down into a per capita figure, Morocco divides its degraded land among its population at 79 sq. m/person, making its ultimate Ecological Footprint for Energy Land .0079 ha/cap.[20]

The third category of Moroccan land use, *Currently Used Land*, includes cropland, managed forests, pastures, and gardens. Morocco's arable cropland totals a meager 9,595,000 ha. Considering factors of importation and exportation of agricultural goods, including cereals, roots and tubers, pulses, vegetables and fruits, tobacco, sugar, managed forests and pastures, the average Moroccan's Ecological Footprint for Currently Used Land breaks down to .6372 ha for cropland, .03 ha for forests, and .09 for pastures, for a total Ecological Footprint of .7572 ha/person.[21]

The final category of land under the Ecological Footprint system of measurements is the *Land of Limited Availability*, which includes untouched forests (which are seen as productive natural ecosystems for their CO_2 absorption capability), and nonproductive areas such as deserts and icecaps. Using the same criteria for measuring desert area as was applied to Spain, Morocco's deserts total some 30,406,100 ha, making a footprint of .81 ha/cap.[22] Morocco's untouched forests comprise a very small 2,509,000 ha, or .067 ha/Moroccan citizen. Lastly, Morocco's polar and alpine territory stands at 446,300 ha, or .01 ha/cap. Added together, the Land of Limited Availability for Morocco is .907 ha/Moroccan citizen.[23]

Adding together all four of our land categories, a portrait of Morocco emerges, which in stark contrast to its northern neighbor Spain, comes very close to (potentially) living within the borders of its own country's territory. While Morocco's actual area is 44,630,000, the land required to provide all of its goods and services is 44,852,040 ha, a difference of 222,040 ha. Morocco's statistics can be misleading, as a great deal of Morocco's land is not ecologically productive and thus unable to meet the consumer needs of its population, while at the same time Morocco remains dependent upon large cereal imports. Nonetheless, *Morocco's ultimate Ecological Footprint is a mere 1.2 ha/Moroccan*, .9 ha less than its Fair Earthshare of 2.1.[24] Morocco's low footprint can be attributed to both negative and positive factors. The poverty of many Moroccan citizens relative to the average Spaniard accounts for a good portion of this discrepancy. At the same time, however, the relatively modest salary of the average

Moroccan and its subsequent buying power has made it impossible for Morocco to achieve the levels of consumption practiced by many more globalized nations. At the same time, Moroccans have had to learn to make do with less, and in many cases they still employ labor-intensive technologies (particularly in the field of agriculture) which have long since been abandoned by Northern countries. Thus, while Spain sees itself in competition with the world, Morocco is forced to focus more effort and creativity on feeding itself before it can turn its attention to world markets. Its Ecological Footprint clearly shows that Morocco is a nation that produces less waste while in some ways more fully capitalizing on its own domestic assests.

A final tally of the Ecological Footprint data is useful in underscoring comparisons between Spain and Morocco:

SPAIN — Actual Land Utilized by Land Type and Expressed in Hectares

Energy	Consumed	Currently Used	Limited Availability
129,158,056	1,969,491.72	50,266,373.37	19,789,900

Hectares of land Spain requires for its level of consumption: 178,320,000
Actual size of Spain in hectares: 49,944,000
Hectares of land in which Spain's consumption
  exceeds its size: 128,376,000
Ecological Footprint of the Average Spaniard: 4.5 ha

MOROCCO — Actual Land Utilized by Land Type and Expressed in Hectares

Energy	Consumed	Currently Used	Limited Availability
15,488,556	295,275.93	39,167,590.051	33,361,550

Hectares of land Morocco requires for its level of
  consumption: 44,852,040
Actual size of Morocco in hectares: 44,630,000
Hectares of land in which Morocco's consumption
  exceeds its size: 222,040
Ecological Footprint of the Average Moroccan: 1.2 ha

The amount of land allotted to every human on earth if we were to all share the Earth's resources equally: 2.1 ha/person

The Conflict Over Land: Ceuta/Sebta and Melilla

What do you think about Ceuta and Melilla?

I don't know. Right now there are a lot of Spanish people there . . . . I don't like borders but they are in their land. I don't know; its not easy. We also have the

Canary Islands, which are like the Falklands . . . . I don't know. It is a delicate situation. I don't like seeing borders in the world; I really don't want to talk about this kind of thing.

Delphina Martin Dominguez, 40, secretary, Seville[25]

What do you think Spain and Morocco should do to improve their relations?

I think relations cannot improve if the Sebta and Melilla problem is not solved. They must give us our land back. We have to fight for it. One fights for two things: his land and his children. If there are not some political relations and negotiations between statesmen in these two countries, the people will retrieve their land by force. If it's up to the people, we will have it back.

Mohammed Nachit, 31, waiter, Fes[26]

The subject of the so-called Two Cities is a source of great pain and embarrassment for the Moroccan people and a source of confusion and sometimes dread for many Spaniards. As it was noted in chapter 3, the city-states of Ceuta and Melilla, which are seated on the Moroccan landmass, became possessions of Spain in 1668 and 1497 respectively.[27] Today, the Two Cities stand as a reminder of a colonial past the Moroccans greatly desire to move beyond, while simultaneously pointing out Spain's willingness to risk contradiction (while it pursues repossession of Gibraltar from the United Kingdom) in the name of maintaining a strategic presence on both sides of the Strait of Gibraltar.

Spanish unwillingness to engage in good faith negotiations for the return of the Two Cities is seen by many on both sides of the Strait as simply a postponement of the inevitable. While there are clear geostrategic and commercial advantages for the Spanish in maintaining Ceuta and Melilla as military bases and trading posts on the Moroccan mainland, their continued foreign possession has cast a pall over Spanish–Moroccan relations. Ecological Realism, as projected by Sustainable Diplomacy, could potentially be an important tool for rectifying this matter, in a manner that promotes both sides' varying efforts to build a long-term working relationship. Let us therefore turn to the ethical norms proposed by the respective Creation narratives of both sides to construct a framework for a new level of cooperation between Morocco and Spain regarding the issue of the Two Cities.

Common Ethical Principles : The Role of the Human and Creation, Spain and Morocco

1. *The human, the Earth, and all of its creatures are in a common, interrelated ontological relationship*

Spain's unwillingness to return Ceuta (or Sebta, as it is known to Moroccans) or Melilla is a clear sign of its ability to compartmentalize its worldview while detaching itself from its responsibilities to its neighbors.

The fact that Spain has never diplomatically acknowledged the contradiction in seeking the return of Gibraltar from the British while maintaining its own colonies only a few kilometers away is a sign of profound denial. Furthermore, from the vantage point of the Ecological Footprint, the keeping of Ceuta and Melilla can be seen as a clear form of ecological colonialism. This fact is only heightened when one considers that Morocco's footprint is only exceeding its borders by 222,040 ha. While Morocco is clearly reliant upon a larger amount of outside lands to feed its people and generate commercial goods and services, Spain's retention of Ceuta and Melilla serves to heighten Morocco's own ecological unsustainability.

At the same time, Spain's retention of Ceuta and Melilla only serves to underscore Spanish ecological unsustainability. As we have seen, a line of unsustainability runs through the last five centuries of Spanish history. This is because appropriating and retaining lands that are not part of its natural landmass indicates Spain's historical reliance upon appropriating other peoples' sustainability in order to maintain its own consumption levels. Yet one must concede that letting go of such patterns is difficult. As the Ecological Footprint has shown, modern Spain's population has become reliant upon a land area over three times the size of its own landmass in order to maintain its collective lifestyle. This being the case, how could giving back the Two Cities be presented to the Spanish people as an advantageous act?

2. The human is in an intimate, evolving, and dialogical relationship with the Creator God. This relationship is shaped by human action and the human's ability to understand the meaning of responsibility to greater Creation

The giving back of Ceuta and Melilla could potentially be a watershed experience for Spain. Such a unilateral act could be seen as an authentic end to the long-lasting era of traditional Spanish colonialism. At the same time, giving back the Two Cities could signal to the Moroccans a real beginning of the end of a long-held Spanish paternalistic attitude toward North Africa. In this respect, ceding these territories could signal the beginning of a new century for Spain, as a country without (contested) colonial possessions. In addition, giving back the Two Cities could lay the groundwork for a new level of respect between Moroccans and Spaniards, a relationship that is currently defined more by economic and military imbalance rather than that of a more equal partnership. In religious terms, Spain's return of Ceuta and Melilla could be seen as a free act of grace by a country that is still inextricably identified with Christianity, toward a nation whose identity is strongly tied to Islam. For this reason, returning the Two Cities would have a profound effect upon the view each country's population holds of the other. These views have been defined by the events that began in 1492, when the Muslim population of Spain was forced to leave or convert to Christianity. In this light, giving back the cities is a modern and concrete way for Spain to show a new respect for Islam and its practitioners in a manner that is meaningful and intelligible to all of Morocco's citizens.

Giving back Ceuta and Melilla could therefore be seen as a Spanish breakthrough in understanding how it is perceived by Morocco, while attesting to its own self-understanding of its responsibility for past transgressions against Moroccan Muslims. While many Spaniards would not identify themselves as Christians, the act of returning the Two Cities would arguably be seen by Moroccans on some level as an honoring of Muslim people by Christians. This act could therefore be a watershed experience, opening the door for new understandings and a new type of alliance between Morocco and Spain.

3. The human is in an intimate and evolving dialogical relationship with human and nonhuman Creation

For many individuals on both sides of the Straits, the Moroccan–Spanish dialogue has for too long been defined by events that took place centuries ago and whose modern meanings are often hidden to those who deny their contemporary potency. The result of nearly eight centuries of Islamic occupation in Spain is more often recognized by modern Spaniards in the architecture of buildings, and overlooked in the architecture of human beings. As we have seen through an eco-historical analysis, Moroccans and Spaniards share common ancestors, and many of those living today in southern Spain share many of the physical characteristics of northern Moroccans. This is a phenomenon that was repeatedly acknowledged by both Spanish and Moroccan interviewees.

Maria José Benítez Rojas

Many people say that there is really no border between Morocco and Spain. Most say that there is another religion, but culturally speaking, physically, you are all the same. And for me, I'm here, and I see a dark haired, dark skinned Spaniard and the face is the same as the faces in Morocco that I know. I am surprised to see that there are not more Spaniards who say "I'm a Moroccan. Yes, I speak Spanish, I live in Europe, but I have a Moroccan mentality. I am happy with this."

Yes but these are concepts that are new. It is very difficult for an Andalucian to feel European. The image of Europe, we are getting it only in the past 5 or 6 years. Until then, southern Andalucia has always been almost Africa. In the north, the Andalucians have always been considered very Moorish, very chauvinist, everything that we attribute to the Moors, that has been attributed to us Andalucians. Now that is changing and the Andalucians couldn't perceive it that way because the only image was the negative image, so how could you identify with that? Obviously, I don't know how an Andalucian can feel European, because he is not the same as a German! Yet an Andalucian has much more in common with a Moroccan. The idea of feeling European . . . well, feeling Spanish . . . we've had that on our shoulders for several years, several centuries, but being Europeans, we haven't had that even for a decade yet. Wanting to be a European now, for us, means being in a welfare state, economic development, and from the economic and welfare perspective, there are a whole set of feelings. Whether we like it or not, we have much more in common with what we have further south, only a few kilometers away. The [temptation] is the need to be European for what it represents.

Maria José Benitez Rojas, 41, psychologist, Seville[28]

Do you think that the Spanish are different from the Moroccans?

Yes, religious beliefs, and even the hierarchy of their society.

What about their physiology?

I don't think that they are different. Probably some of their features are—especially the people in the north of Spain.

And as for the (Spanish) people in the south?

There are similarities, like us in the north of Morocco—there are many (of us) who look like Spaniards. But these differences could be found even inside one country from one region to another.

Tijani Aharchi, 38, nurseryman, Fes[29]

A people and their land cannot be separated, according to Daniel Spencer in his explication of Ecological Location. The people and the land share a dynamic relationship that defines its meaning and character. When the lands of what appear to be two people share the same ecosystem, this compels us to reassess the means by which we have separated the people. Spain must therefore be compelled to reassess its continued possession of Ceuta and Melilla. Ecologically speaking, the ecosystems of the south of Spain and the north of Morocco are part of a continuum. For this reason, it can be argued that the land areas that comprise the closest points of physical contact between Morocco and Spain are expressions of ecological, social, and physiological limenality. Out of this limenality new languages and new religious understandings can most easily be formed and practiced.

Yet faith and a common reading of the land is not an adequate basis upon which to build long-lasting diplomatic ties. This is why the work of Sustainable Diplomacy in our case study also focuses on the need for both sides to acknowledge what history and modern science have proven to be true: that Spain and Morocco have a common genetic ancestry, and live in a common ecosystem. Advocates of Sustainable Diplomacy must therefore invite both sides to read from their common historical, ecological, and religious texts, and must promote a type of reconciliation that acknowledges that the Spanish–Moroccan conflict is based in part on age-old religiously founded misinterpretations and fears. One lesson to be learned is this: while history can shape our modern circumstances and our understanding of ourselves and our neighbors, it is not the only way to interpret and understand contemporary events and opportunities for change. The relationships between bioregions, nation-states, humans and nonhuman Creation are dynamic and evolving, and therefore are always in need of individuals and groups who are willing to engage them in their most contemporary manifestations. The tools provided by an Ecological Footprint analysis can be exceedingly helpful in making these conversations possible. In turn, by Spain relinquishing Ceuta and Melilla, this new type of conversation between Moroccans and Spaniards might more clearly and quickly emerge.

4. The human is in possession of tremendous power vis-à-vis Creation, bestowed by God's very design

In the course of this text, many different types of power have been named. Among these powers are the power to become sustainable, the power to control one's level of consumption, the power to reach a point of ecological equilibrium, and the ability to move toward a bioregionalistic worldview and a new understanding of sovereignty. Yet there are other powers of equal importance, including the power to admit wrong and the power to forgive. These are all powers that Spain and Morocco must be invited to wield together.

Ceuta and Melilla have stood for centuries as a reminder to the Moroccans of their military and economic inferiority to the Europeans. In the wake of the Inquisition's banishment of Muslims and Jews from the Spanish mainland, Ceuta and Melilla stand as a concrete reminder of Spain's continuing ability to control the lives of ordinary Moroccans. In this respect, the Two Cities' continued possession by Spain comprise an open wound from which many Moroccans are unable to find relief. Thus the power implied by Spain's ability to retain these possessions is an outward tangible sign of Spain's unwillingness to repudiate its past as a colonial power, and its continuing unwillingness to be open to a profoundly new relationship with Morocco and its population.

The time has now come for Spain and Morocco to instigate a lasting reconciliation by returning the Two Cities. In order for this to come to fruition, each side must be willing to unilaterally offer an admission of transgression (in the case of Spain), and the willingness to forgive (in the case of Morocco). Both these acts are clearly within the power of both nations, and both sides have something tangible and valuable to gain. A new level of cooperation across the Strait is exactly what will be required in order to resolve the other two conflicts this book will address: that of the "fish wars" and the continuing conflict over illegal immigration. The resolution of any or all of these conflicts depends upon the judicious use of human power at the state level and at the individual levels.

5. There is a language within and expressed by Divine Creation

The text that is Creation is in grave danger, because very few people have shown any interest in reading it. As the Ecological Footprint has shown, both Spain and Morocco have been unable to live within the means provided by the land within their own borders, and they are therefore physically dependent upon the consumption of products that originate outside of their respective countries. Yet this is really just the beginning of a much larger problem. The generation of CO_2 by both countries requires carbon sinks that neither side is ecologically able to provide. Working as separate entities, Spain and Morocco have shown that they have limited canvases to work upon for change. However, working together as a unified bioregion, these two countries have a greater opportunity to diminish their collective Ecological Footprint.

The fact that Morocco's footprint is appreciably smaller than Spain's is in many ways due to Morocco's relative poverty in comparison with its northern neighbor. Yet the term "poverty" itself should be treated with some degree of suspicion. This is because in some cases, poverty is simply equated with a lack of what are thought to be "proper" developed-nation levels of consumption. While it is true that Morocco has a much higher level of unemployment than Spain, has a lower average wage, and is dependent on foreign nations' donations of grain in order to feed itself, Morocco is arguably the repository of critical ecological knowledge, which has been almost completely discarded by Spain. Such knowledge is embodied in the ability of the majority of Moroccans to live simply, to get more from less, and to survive without many amenities that those in the most developed nations take for granted. Many more Moroccans than Spaniards possess knowledge of how to raise their own crops, through firsthand experience or the experience of their parents.[30] Spaniards, who only 20 years ago lived at a development level much closer to Morocco's, have been made members of "the developed nation's club" through Spain's recent membership in the EU. Spain's consumption levels have subsequently become much higher, as their per capita Earthshare has grown exponentially.[31]

The language of Creation is a language that is most easily spoken and understood by those who live close to the land. The rural flight to urbanized areas is a phenomenon known to both Spain and Morocco. Yet in comparison to Spain, many of Morocco's urban inhabitants have moved themselves and their families to cities much more recently, while leaving a higher percentage of the population behind in rural areas.[32] Thus, in comparison to its northern neighbor, it could be argued that Moroccans are in greater possession of a type of knowledge and experience that can only be gained by living closely with the land. Such people know where their food comes from and are quick to explain how it is produced. On the other hand, more and more Spaniards, due to their urban locations, are unable to name where their food comes from, or how it was produced. Such a population is by default, agriculturally, largely a consumer society.

It would appear, therefore, that it will be left to the Moroccans to help teach Spaniards how to live more simply, with fewer amenities, and with the ability to stretch fewer materials further. This is a gift that a Morocco which has been freed from the embarrassment and humiliation of Spain's possession of the Two Cities might give to Spain. Each side clearly has something precious to offer the other. The situation is filled with opportunity for the practitioners of Sustainable Diplomacy.

Points of Disagreement: Naming, Sin, and the
Role of the Human in Spain and Morocco

While the points of commonality offered by the above framework provide one avenue of negotiation for the practitioner of Sustainable Diplomacy, one cannot lose sight of the points of contention between Morocco and Spain regarding Ceuta and Melilla. These points of contention must be kept in tension with the framework thus far presented, as they provide a critical key to understanding what must be overcome if Morocco and Spain are to eventually cooperate regarding the Two Cities. Let us therefore return to the points of disagreement regarding the Creation stories of Islam and Christianity to guide our examination.

Naming (Human Power and the Will of God)

As Spain moves toward closer integration with Europe through its membership in the EU, Spain and the Spanish peoples' attention is being drawn northward. At the same time, Spain is being invited to reinvent itself as a fully European nation. For many years, northern Europeans told the joke that the boundaries of Europe "ended at the Pyraneés," implying, pejoratively, that Spain and Portugal were somehow more African than European. While northern European chauvinism with regard to Spain and Portugal has not ended with the Iberian Peninsula's EU membership, western European integration has afforded Spain and its citizens the power to name and rename themselves and those around them. Spain therefore has named its northern neighbors as the primary source for its political cues, and is arguably the most enthusiastic member of the EU, its membership coming after years of isolation during the Franco period. Spain names itself now more than ever as a European country with European ties, at times de-emphasizing what some have called Spain's liminal role between Europe and Africa. The degree to which Spain's African heritage is lost in this transition remains to be objectively measured. One thing, however, is clear—the EU has made Spain its gatekeeper, and charged it with keeping illegal African immigrants out of the EU. This is but one price that Spain has been forced to pay in order to call itself fully "European." Spain must name on one level one of its closest neighbors, Morocco, as an adversary, as it names its newly minted northern neighbors "allies." It is in this milieu that Spain must grapple with the questionable nature of its continued possession of Ceuta and Melilla, and not be completely seduced by its new recognition as a full member of one of the world's most powerful economic blocs.

While Spain enjoys the fruits and challenges of EU membership, Morocco languishes economically. With soaring unemployment for a burgeoning younger generation, many younger Moroccans despair at ever finding meaningful and sustainable employment. The most desperate risk their lives in small boats in an effort to cross the Strait and benefit from what Moroccans see as their clearly more economically advantaged Northern neighbors. Many Moroccans, when questioned, will name the will of God as being at the root of their desperate circumstances.[33] Others name Europe itself as "Fortress Europe;" a group of nations bent on retaining their blessings for their own benefit, including the city-states of Ceuta and Melilla. Some will enjoin this statement by pointing out the blessings that God has rained down upon their European neighbors, in the form of more rain, more employment, more mobility, more wealth, and more opportunities. As Morocco comes to grips with these disparities, it does so as a Muslim nation that believes that it was originally called to lead other nations, as in many cases it has within Africa. Yet for many Moroccans Europe retains the title of "other." As Spain seeks to further integrate itself into Europe, it is viewed by Moroccans as making the lives of many of their Southern neighbors more difficult. In turn, Moroccans have been given many names by Spaniards, and few of them are complimentary.

Sin (Historical and Contemporary)

The transgressions that have taken place between Spain and Morocco over the years can be described in a number of terms. While the word "sin" is not a favored word among diplomats, it is arguably descriptive of actions taken by both sides. How sin is identified, named, and responded to depends on who is doing the naming, and who is responding to the transgressive actions taken.

In the eyes of Spaniards, the sin of colonialism largely lies in the past, in actions ascribed more often to ancestors than to living people. Any gains that Spain accrued from its colonial past are understood by many to have been long absorbed by history. Such history is now being placed even further from the collective Spanish memory by Spain's new role and acceptance in the context of its membership in the EU. Freed from the embarrassment of Francoism, Spain moves further and further away from seeing itself as a contemporary aggressor. In this respect, dwelling on past events are often relegated to being seen as a historical exercise. Any doubts many Spaniards have regarding this phenomenon are diminished by the gargantuan shadow of modern American economic and military colonialism.

For Moroccans, however, the sin of colonialism is real and contemporary in the holding of Ceuta and Melilla; and is thus projected upon the modern generation of Spaniards. While a number of Moroccans will concede that it was the Maghreb that first colonized Spain, the negative consequences of these actions are seen as being eclipsed by modern Spanish policy toward Morocco and the Two Cities. As in Qur'anic interpretation, more recent developments have abrogated older events in the eyes of the Moroccan populous. For most Moroccans, just as for the majority of Spaniards, responsibility for the sin of colonialism lies with ancestors and not with the living.

Role of the Human (in Spain and Morocco)

As has been noted, Spain's new role among western European nations has brought with it the beginnings of a Spanish self-reappraisal. Now that Spain finds itself taking what it considers to be its long-denied proper role among European nations, it is more intent upon Northern, not Southern reintegration. No longer cut off from the rest of the continent by the legacy of Franco, or even earlier manifestations of northern European chauvinism, many Spaniards sense a tremendous new power through EU membership, a common currency and full participation in NATO. In these new alliances, Spaniards have a variety of new avenues for denying their historical and genetic ties to North Africa. In this respect, Spain is in a sense being offered the chance to become once again "White." If this is the path Spain and its citizens ultimately choose, the prospects for a meaningful rapprochement with Morocco will significantly diminish.

While their neighbors to the north prosper, the Moroccans are left with being on the losing side of globalization, with higher unemployment, and an emerging pariah status in Europe.[34] For these reasons and many others, the full Muslim role of "vice-regent" for the first human is one that has escaped the grasp of the average modern Moroccan. Instead, Morocco has been forced to turn inward, in an effort to confront long-standing internal problems, including a crumbling infrastructure, an inability to feed itself, unequal education systems, and enormous disparities of wealth. In the midst of these struggles, satellite television and the Internet make Moroccans highly aware of how people live in more prosperous nations. These new contacts will almost certainly be accompanied by new desires and disappointments for the Moroccan people.

The role of the human is rapidly changing and enlarging for Moroccans and for Spaniards. The manner in which both sides negotiate these changing roles will have profound and long-lasting consequences for human and nonhuman Creation. As we turn now to the topic of natural resources, let us not forget the multiple roles inhabited on both sides of the Strait, as we continue to seek common ground despite any differences in perception, wealth, or belief.

THE CONFLICT OVER NATURAL RESOURCES: THE TREE OF LIFE AND THE TREE OF BEING, THE CONSUMPTION OF NATURAL RESOURCES, AND THE FISH WARS

The practice of Sustainable Diplomacy will be determined by the ability of its proponents to engage people on the ground across a broad section of the population. Disputes and pacts between national governments often do not reflect the concerns of the populations they represent. For this reason, one of the principal goals of Sustainable Diplomacy is to move from the established vehicle of individual diplomacy to the practice of systemic diplomacy, which embraces dialogues, peoples, organizations, and concerns that are often ignored or marginalized at the national level. Sustainable Diplomacy therefore envisions a greater role for NGOs, for those who speak with a religious voice, and for many other groups and individuals who work to forge connections across barriers that are only now being named.

The task ahead will place great value on the ability of people to listen carefully before they speak and to listen to voices that have been relegated to the margins. We must return to a basic knowledge of how people live: where they find their food, how they manage to provide themselves with shelter, and what beliefs guide the pursuit of their goals.

Understanding Patterns of Consumption: Daniel Palau i Valero and Muhammad Amouche

The interviewees presented here seem at first glance to be poles apart. The first is a 27-year-old Catalan seminarian from Barcelona named Daniel Palau i Valero. The second, Mohammed Amouche, is a 78-year-old Imazighn retired factory worker and farmer from a tiny village near the edge of the Sahara. Despite their differences, however, both are guided by a deep faith in God and an understanding of their own intrinsic connection to the land on which they live. In seeking an end to the ongoing dispute between Morocco and Spain regarding fishing grounds, the knowledge presented by the following individuals will be critical in understanding the points of commonality and difference between Spanish and Moroccan patterns of consumption. At the same time, we must listen to these and other voices for wisdom that could one day be employed in constructing treaties whereby the Moroccans and Spaniards could share their natural resources with each other.

Daniel Palau i Valero

Do you grow the food that you eat?

No.

So where does your food come from?

I suppose from the multinationals and very little from the land. From the supermarket.

When you are in the supermarket, do you have an idea where some things come from?

I don't buy many vegetables, because I normally buy them in [my] village [since] I have that possibility, so . . . If I walk along the vegetable section, I don't ask myself where they come from. I know they either come from France, or from here, because they are vegetables that I see being grown in my village. Tomatoes, lettuce, anyway . . .

Do you know how to grow food or raise animals?

Animals no. But if I were in the countryside, I wouldn't have any problems grow[ing many things], because my grandfather was (you are lucky with this question!) a farmer. So I went to the land with him and saw how he worked the land—potatoes, carrots, lettuce, tomatoes . . .

So can we say that your grandfather was the last person in your family who did this kind of work on a regular basis?

Yes.

In your opinion, what is the most important animal, plant, or tree in your region?

The most typical animal, here in Catalonia (because I'm talking about the land, I talk about Catalonia) . . . well, many and at the same time none, because maybe I could talk about a cow, about a wild boar, which is a wild pig, maybe I'll stick to that one. Yes. The plant, maybe a rose, because they have many different meanings here in Catalonia. The 23 of April is the day of San Jordi and the children give their beloved a rose, or their supposed beloved. And the girl gives the boy a book. So the rose has the longest tradition here in Catalonia. A tree, . . . well . . . a pine tree, because it is the tree of the mountains and there is one tree in Catalonia, that is one with three branches, that is something special.

What is your responsibility in helping your next generation?

Look. Most of all, [to] give my testimony and not lose the history of our land, our country, and let the next generations know, so that they want to love their history more, the history of Catalonia, the history of their family, their personal history, and to let them get to know themselves better.

Is the health of the land connected to the future of the next generation?

Surely. In several ways. Because when natural resources are gone, it will surely affect future generations and when transgenic [genetically altered] food has become more popular than natural food, it will surely affect metabolism and the way that people function. I think it would be an important loss to forget about the products from the land, natural products that God gives us in his Creation. It's a pity that multinationals give priority to economic progress and economic income, instead of history and the natural products of a country.

So what are you going to teach the young people in your job about Creation?

What comes to my mind now, is, that Creation is a gift of God that we have very near, that is a gift to love, that we should respect, that the Earth is a place for every-body, that, although we don't relate well, the Earth is not the possession of a company or a multinational, nor of a farmer, but it's the history of humankind. And it is that through Creation we can also see, thanks to faith, its testimony. It wants to say something, not only through people, nor through strange spiritual experiences. Starting from life and such natural things like a tree, a plant, a flower, a garden . . .

What does your theology teach you about your connection to the land and other living creatures?

That is a subject that I have to do this year. I think that, what Christian or Catholic theology wants to teach, is this: Creation is a sign of God, humankind is a reflection of God's love, and everything leads us to God somehow, although we go backward, burning forests to get economic benefits, mak[ing] war to get a hand's span of land. I think what it teaches is this and that Creation is this, a gift.

Do you have special responsibilities regarding Creation because of your theology?

Surely, of course. It's responsibility with everything and with myself. I am also part of this Creation and also life, which my parents have given to me, but to them their parents and so in the origin of life and Creation, although many people don't want to see it, but there is God. And even if it were just a little spark of love, well, that little spark is God. As far as responsibilities, well, love yourself, respect the life of the other, respect nature, respect relationships and know how to take advantage of it and live it in a happy way.

Is there a specific way in which you understand God's work in Creation?

There is a sign in nature that speaks to me about God, which is the fog. Because the fog is like God: it envelopes you and catches you and moistens your skin and it's like God who envelopes you and is all around in your life and relationships and who gives you confidence and faith to continue walking a few meters ahead. For me. The moon at night, for example, or the stars. It's very typical when I go on an excursion with the scouts, you look at the sky and the stars and the moon. When I see the moon, it's . . . the moon from the night when Jesus died. And it's the same moon that Jesus saw and that I see.

One more question about Christianity in Spain. I am a foreigner and I don't completely understand what happened in Spain. I think there are many people in Spain in their thirties and forties who don't like the church, because they equate it with Franco and fascism. Very often here in Spain, when I say I am a Christian, many people think I am very conservative, but in fact I am progressive. What do you think is the future of the church in Spain and is it possible to teach people here that there are many possible political positions, while being a Christian?

It happens, that is true. Look here, in the Franco era, . . . I have told you before about the political matters between Catalonia and Spain. As you can imagine, I haven't got anything to do with Spain, very little, although my grandparents from my mother's side were from Andalucia. No problem. But here, during the Franco era, there was a strong oppression toward Catalonia, especially toward the language and culture, in such a way that many people had to emigrate, most of all intellectuals and people from the church. What happened? Little by little pro-state bishops arrived. It's a historical footprint that we have to live with and know that that period is finished.

But this era now has other characteristics and within Christianity there are many positions. We Christians are more authentic every day; we have to vote [for] a political party, because we have an obligation. You cannot [succumb to] your inhibitions and forget about society, you have to vote. But as a Christian you transmit a shimmer and will to collaborate, each one from different positions. One can be from the left, another more right wing, but if they are authentically Christian, they will complement each other. This is the theory, in reality everybody fights with everybody. Sometimes you are discouraged and sometimes you say it isn't right. They are young people who, without forgetting history, want to give the testimony of common sense, of living together in peace, who want to grow inside. There are people that want to get to know the heart of others and get closer to God and build a fair society and one without discrimination of sex nor religion, because we don't want that either. But this is a slow process and evil is always seen before these little signs of good relationship [can arrive].

Political parties can do very little, in economics almost nothing, especially small countries, because the USA is very powerful and the oil-producing countries [are] as well, and Spain and Europe are quite limited. And a left- or right-wing government doesn't make any difference, because they are very limited. They don't have such potential as to say: "we stop trade with this or that country." Finally everything ends up with getting to know a personal testimony of someone that you know and through which you create a network of positive relationships and relationships in which you don't forget history. You don't have to justify your position every time. Let Jesus and the angels reign. This is the theory, you know that reality is different, because there are young people [who] want to commit themselves to a young church, that has nothing to do with Franco, because they don't like [him].

Now we are not like we were before. In the end you will say that it doesn't make any difference to you . . . If you want to get to know me, you will also get to know a part of the church. If not, it's you who is closing yourself up in your world, in your ideas. In the end you will also have to do a bit of mental hygiene and say that the church also apologizes. And which multinational or world bank has apologized for charging interest to a poor country? The church has done that and you don't realize it. Jesus is in the center of the church and Jesus is that spark of love. And sometimes we have to say: "hello there, relax." [We need] faith, calm and tranquility, and [we need to] say: "look, if you want, we can talk, but I am not going to solve any of your problems, because I can't and because I don't want to. Solve it yourself. If you want, we can talk and share your worries and desires. I also have mine. I am also on that road." People will have to learn how to live in peace and find God in their lives and transmit that to their work, their family. They are all signs of God. You might [struggle with] depression or die from AIDS, or be unemployed, or at the verge of divorce, or your son is mentally handicapped. [Within all these circumstances] there are calls from God. In depressions he says he's on your side. In happiness as well, when you have found a girlfriend, when you have passed an exam, doing your thesis that costs a lot, knowing what suffering means, away from your friends and family. God wants to tell you something, doesn't he?

Sometimes, when things happen to me, I ask myself why, why? Because they love you and that's it. All this is a learning process. Faith is not a magic pill. All these experiences will have to settle in your heart and you should trust in them. You need to be able to talk to a group, a group . . . with the gospel, which in the end is a criterion that [prepares Christians] for everything. Jesus also tells different things to each of us through the gospel, or that children can talk to their parents or people in the villages with their priest. This is the theory. Life is, that, if you love people or not, if you have a transparent heart ...

Daniel Palau i Valero, 27, seminarian, Barcelona[1]

Muhammad Amouche

Why do you choose to live here?

My father left me here. My father had cattle. I worked here and there in Tasourt to support myself. And I worked also with my Uncle. We bought dates. At that time there were only camels and mules. [Eventually] people cultivated and sowed with their cattle. They lived and settled in peace. Times became difficult but people were not hungry.

What does life mean for you here, living in the country?

This land, and all these ethnic groups around Tafraout, here their people live. There are poor people, middle class people, and well-off people. They cultivate, sow, and farm. Those who move away from this land come back to it, because no other place beside your land will accept you.

What do you like in your life?

I want to spend my life in religion, prayers, ablutions, and support my children if they want me until I get them, males and females, married and till I go to my place, which is the grave. I don't wish to do commerce or something else. I don't fast. I'm with my children until I pass away.

Outside your home where do you spend your time?

The mosques and souks. I do shopping in Tafraout in the Thursday market [weekly souk], Tiznit, the Friday market, and go to places of dates and "Fomlahcen."

What's the best thing in this place?

People in this place are religious. They do their prayers. They go to the mosque. They take care of their animals. Their cattle are their life. They are religious. They are people who care for their land, others do commerce, and others go to Europe. God help them all.

Do you raise the food that you eat?

We raise grain, wheat, barley, and farina. There are farmers here who collect grain when it is abundant. But in these last four years of drought there was no water and no hay. They get milk from their animals and do herding. This is life in this place. People who have money get their needs met, and those who don't have to sell their cattle. And there are poor people too.

Where does the food you eat come from?

It comes from souks. I buy flour, farina, wheat, whole wheat flour, vegetables, and fruits because I am no longer able to cultivate land.

Do you know how to raise vegetables and animals?

Yes, I do.

Who taught you that?

I learned it myself. I found my parents, my uncles in this place, this job they had. People learn things by themselves. We cultivated using our animals, but now we use a tractor to cultivate our land.

In case you did not know how to do these things, who can do them for you?

All people can do that. They know how to sow grain, cultivate soil, and mow. They know how to collect grain and so on.

Outside your family, who are the important people in your life?

All people are good. There are no people here who shout at you. They are religious. They respect themselves. People who go astray are given advice. The place is like the place of Afqqa. Afqqa is a place of religion and schools. People care for their belongings and commerce and they work, they sow, they reap. But people can no longer do this . . . because there is no rain.

What sort of relationship exists between yourself and the people?

I found that my parents were like this, and we are like this, and we hope that our children will be like this. We don't show our children dishonest ways: stealing, gambling, wine. All the religion is like this. The only way here is religion. This is how people live, and children for those who have them. "Where you put your head you find it." You now, you support your parents. Children, if they are made good by God they can support and work for you. If they don't then there is but God. This place is like this and the people of this region are good. They're religious. They cannot do bad to you. They do not poison you or poison your children. You can be religious, do your prayers. There are thinkers. This is a country of learning. The Zaouia down there is where Mokhtar Soussi was buried, may he rest in peace. This is his country. He is from the Zaouia of Darkkawa. People came from everywhere in Morocco to the Zaouia by bus. The Chief of the Zaouia, Al-Hajj Abdessalam is in Agadir. He is in "Tamchiekht" (a place for the sheik and his family). His brothers are here. They are called Darkkawa. There is the place of religion. Their focus is only religion and their prayer beads.

What relates people to each other in this place?

For example we sit together. We talk about surviving. We laugh. We talk about agriculture. We ask God to give us rain. We ask God to help our King and protect him. We pray for Moulay El Hassan, may he rest in peace. And we ask God to bring the Sahara back to Morocco. We ask God to help the Palestinians get their country back. We talk about countries of religion and Al-Hajj. We put faith in our King, may God help him and his ministers.

Which animals are the most important here?

There are goats and sheep. This region supports a lot of them. His Majesty the King, may God help him, and may Moulay Al Hassan rest in peace. This country always says "yes."

What is the crucial animal or tree here?

You're looking at it now (gesturing to the table that is being prepared for dinner). Oil or Almonds.

Olive oil?

Yes, Olive oil.

What is your feeling in relation to the land you live on?

Calm. We sleep, drink, and eat peacefully. There is nothing that can get us nervous in this place. The King endows us with awe that does not exist in any other country, God help him and endow him with Baraka, and may God help Moulay Rachid.

What is the origin of the relationship that connects you to the land? Do you own any land?

We have land left to us by our grandparents and parents. We cultivate it. It is hard to leave this land, even in times of drought. People here survive and the King, may God help him, takes care of people and provides them with water. He is building roads for them and supplying them with phones, electricity and people are satisfied. Our country here is peaceful. And the whole country from Tangiers to La Gouira [the most southern point in the Western Sahara] lives in peace. May God help his Majesty the King and endow him with Baraka.

Is your work related to the land?

I am a farmer, and my children are farmers. If my children want to follow my advice this place is for agriculture. If my children have a job like you they have to follow their children, and have their children support them. The land of Morocco and that is all. Our country from Tangiers to La Gouira is in peace. Nobody knocks on your door in the night. We have peace in our country. There is a Prefecture that supplies guards, municipal guards. There is a governor who takes care of these regions. May God help our King who provides water for these regions. God gives us flour and His abundance. We have hay, barley, and the souks are full. His Majesty the King gives everything to these regions, may God help him.

How is your life affected if the rains don't come?

If there is no rain I am one of the people. If there is no water we ask God to solve the problem. Water is everything: "We made from water everything that is alive" [Quoting the Qur'an]. If there is no water you cannot do your ablutions. If water does not come first during the meal you cannot eat. People die on the spot for lack of water, but you cannot die on the spot for lack of food. If there is no water you have to ask God to provide you with water. If there is water you cultivate, you mow, and there is barley, farina and other things. May God help the King provide us with everything.

Does your religion or culture oblige you to protect land or creatures?

Yes, for I can't do any harm to anyone's land, or bear false witness, or wrong my brother, or my neighbors, or the public or citizens. I am a straight citizen.

What are the obligations that force you to behave so?

They are: to adjust yourself to the discerned path of religion. All these things are between my God and my grave. For when you die, everything is left behind apart from your deeds. You have no longer any lawyer to defend you, you have but what you did. If you falsified, or did injustice, may God help people [who] are like this.
Muhammad Amouche, 78, retired factory worker/farmer,
Ait Abdellah Tahjala, Ait Oufka Tafrout[2]

The work of Sustainable Diplomacy demands that we try and imagine Daniel Palau i Valero and Muhammad Amouche in direct face-to-face conversation with one another. Clearly, their respective faiths and their abilities to see God in the land and its inhabitants are two points that bind them together. They both view the land and its abundance as a Divine gift. Both, as well, fear for the future. While Daniel fears the introduction of genetically altered food and its possible consequences for the human body, Muhammad decries the lack of rain and how it is starkly altering the way of life of the people of his region. Both are dependent on the commerce of natural resources, and yet each has a different level of knowledge of the means of producing such resources. While neither Daniel or Muhammad have met in person, their stories—told together side by side—demonstrate that the land is a good text for Spaniards and Moroccans to begin their dialogue. Through this common hermeneutic, they can begin to concretely describe human experience in the human and nonhuman world.

Sustainable Diplomacy must put members of respective populations who are often separated by distance, class, race, age, language, religion, gender, and opportunities into conversation. The Sustainable Diplomat is therefore someone who can both listen to and record people's stories and, at the same time, be willing to invite both sides to recognize their points of common intersection. This is an exercise done not simply to exchange information, but also to build avenues of communication that could be later used to undergird physical cooperation between two populations.

Yet, as it has already been presented, while the land is the first text that practitioners of Sustainable Diplomacy should read, the common lessons learned from the land must inform a second reading—one that comes out of an exchange of sacred stories from each side's religious texts or traditions. These sacred texts and stories should be read through the hermeneutic of the land as it exists today, never separating the human and nonhuman occupants from the land itself. Such a manner of reading sacred stories and traditions should render new interpretations and understandings that speak directly to the circumstances at hand. At the same time, these interpretations should be accessible to all who wish to hear them and add to them. In this way, the stories remain alive and dynamic, like the land and the people and the God they describe.

Reading the Sacred Texts and Traditions: The Tree of Life and the Tree of Being

The Biblical Tree of Life and the Tree of Being as it is known to Muslims provide a second shared text, which Christians and Muslims can use to build common vocabularies and common bridges between their two traditions. Read as a story through the hermeneutic of the land, the sacred trees of Islam and Christianity afford rich opportunities to explore how each tradition talks about Creation, its natural resources, and human responsibilities. Through examining the stories of these Trees, we can begin to examine Christian and Muslim attitudes toward consumption, the gifts that are natural resources, and the continual human struggle to choose the correct path in sharing and maintaining the limited capacity of earthly Creation.

The Tree of Life and the Tree of Being share many things in common. Among these similarities is the limited amount of space each garners in its respective sacred text. It is for this reason that a consideration of the trees must include extra-Biblical and extra-Qur'anic commentary in order to flesh out their meanings and roles for Muslims and Christians. There is no lack of such commentary from the Christian perspective. For the purposes of this examination, the focus will be on the work of the Biblical scholar Pablo Richard. As a twentieth-century liberationist scholar, Richard offers an expansive and challenging reading of the role of the Tree of Life, and an invitation to push its role further than many of his colleagues. From a Muslim perspective, among the only known commentaries on the Tree of Being was one attributed to the celebrated twelfth-century Andalucian Sufi theologian Ibn al-'Arabi, in his work entitled "Shajarat al-Kawn." Because the Tree is mentioned only once in the Qur'an, texts from Ibn al-'Arabi's work will be used as a supplement that is representative of an important branch of Muslim tradition.

Many different qualities have been ascribed to our two Trees. Both Muslim and Christian traditions consider the Tree to be the *axis mundi* of their respective descriptions of earthly Creation. While Genesis ascribes the capacity to make humans immortal to the Tree of Life, the Qur'anic sura An-Nur describes the Tree and its Creator as a source of great wisdom and light. Interestingly, the motif of light shows up as well in Revelation, where the presentation of the Tree of Life is accompanied by a commentary that tells us that there will be no more night, and thus no need of the sun or lamps. God, it is said, will provide all the light that is needed.

From the Qur'an and from Muslim Tradition

God is the light of the heavens and the earth. The semblance of [God's] light is that of a niche in which is a lamp, the flame within a glass, the glass a glittering star as it were, lit with the oil of a blessed tree, the olive, neither of the East nor of the West, whose oil appears to light up even though the fire touches it not,—light upon light. God guides to [God's] light whom [God] will. So does God advance precepts of wisdom for [humans], for God has knowledge of every thing.

—An-Nur 24:35

The shoots of this tree differ, and its fruits are of different kinds, that there may appear to the sinner the mystery of [God's] forgiveness, of [God's] mercy towards those who do good, [God's] kindness to the one who is obedient, [God's] favor to the believer and [God's] vengeance on the unbeliever. In [God's] existence [God] is far separated in holy separation from contact with that which [God] brought into existence, from being near it whether joined to it or separated from it, for [God] was when there was no universe, and [God] is now as [God] was then, neither joined to nor separated from the universe, because being joined to and separated from are attributes of things that have come into existence, not attributes of the pre-existent One.

—*Shajarat al-Kawn*, Ibn al-'Arabi[3]

When the trunk of this tree was firmly fixed and its branches also stood firm, it pushed outwards its edges toward thy Lord till its one side overtook the other and its end reached to its beginning, for since the beginning is "Be!" (*kun*) its end is "will be" (*yakun*). Thus however many you may count its branches to be, and no matter of how many kinds of seeds, yet, its root is one, namely the seed of the word *kun*. This is that were you to gaze with intent gaze of your sight, you would see the shoots of the [celestial] tree TUBA clinging to the shoots of the [infernal] tree az-ZAQQUM, the coolness of the [celestial] breeze al-QARAB mingling with the heat of [the violent, hot, infernal winds] of as-SAMUM, and the shadow of the heaven[ly shade] al-WASL joined by the shadow of [the infernal] YAHUM. Each attains his allotted share of fortune, so that one drinks from a cup that has been sealed, another from a cup that has been blackened, while another between them is forbidden to drink at all. So when the progeny of what has been brought into existence appear from the presence of non-existence, there blow upon them the breezes of Power, there flow towards them the gentle winds of Wisdom, and rain upon them the clouds of Will, with all the wonders of creativeness, so each shoot thereof sprouts forth with what has been ordained for it from of old, and in its constituent parts are combined both health and sickness.

—*Shajarat al-Kawn*, Ibn al-'Arabi[4]

From the Bible

Then the Lord God said, "See, the [human] has become like one of us, knowing good and evil; and now [the human] might reach out [a] hand and take also from the tree of life, and eat, and live forever"—therefore the Lord God sent [the human] forth from the garden of Eden, to till the ground from which [the human] was taken. [God] drove out [the human]; and at the east of the garden of Eden [God] placed the cherubim, and a sword flaming and turning to guard the way to the tree of life.

—Genesis 3:22–24

Upon my bed this is what I saw; there was a tree at the center of the earth, and its height was great. The tree grew great and strong, its top reached to the heaven, and it was visible to the ends of the whole earth. Its foliage was beautiful, and its fruit abundant, and it provided food for all. The animals of the field found shade under it, the birds of the air nested in its branches, and from it all living things were fed.

—Daniel 4:10–12

Then the angel showed me the river of the water of life, bright as crystal, flowing from the throne of God and of the Lamb through the middle of the street of the city. On either side of the river is the tree of life with its twelve kinds of fruit, producing its fruit each month; and the leaves of the tree are for the healing of the nations. Nothing accursed will be found there any more. But the throne of God and of the Lamb will be in it, and [God's] servants will worship [God]; they will see [God's] face, and [God's] name will be on their foreheads. And there will be no more night; they need no light of lamp or sun, for the Lord God will be their light, and they will reign for ever and ever.

—Revelation 22:1–5

Common Ethical Principles: The Lessons of the Tree of Life and the Tree of Being (Shajarat al-Kawn) in the Biblical and Qur'anic Accounts

1. We live precariously and temporarily between heaven and hell, both of which can be reflected in earthly life depending on the path of human action

The Biblical Tree of Life and the Qur'anic Tree of Being stand as a reminder to their human onlookers that human earthly lives are temporary, while, in contrast, the Creator God is eternal. At the same time, both Trees stand as signs of Divinely imposed human limits. For while the first couple ate from what is known in the Bible as the Tree of the Knowledge of Good and Evil, the humans of Genesis were ejected from the garden before they could come near the Tree of Life.[5] In a similar fashion, Ibn al-'Arabi speaks of the Tree of Being as standing at the root of God's Creation; its continued existence is central in explaining the nature of the universe while reminding the human of the true, eternal treasure of God.[6]

Both the Muslim and Christian traditions attach critical eschatological significance to both the Tree of Life and the Tree of Being. As a bookend of the Bible in the Christian tradition, the Tree of Life, first presented in Genesis as a means for obtaining eternal life, returns in Revelation as the giver of life to the inhabitants of the new heaven and Earth, or the New Jerusalem. For Pablo Richard, the emergence of the New Jerusalem is not in fact an end of earthly history but the beginning of a new history, one in which those who have hunger are fed, and those who have been excluded are included.[7] Muslim tradition as interpreted by Ibn al-'Arabi implies that the fruit of the Tree of Being itself brings of mercy and forgiveness to those who sincerely follow God, and metes out God's vengeance to the unbeliever.[8] Both Trees, therefore, reflect the power and scope of human choice and God's ability to include or exclude. In the Christian tradition, the Tree of Life is ultimately a gift reserved for those in heaven, while for the Muslims, the Tree of Being participates in a different kind of separation, appearing to both the heaven-bound and the damned in completely different forms, its fruit gives out rewards and punishments as merited.

2. *There is an* axis mundi *in God's Creation, providing balance, wisdom, cohesion, and logic to the biosphere. It can be seen and understood by those who look for it and strive to interpret it*

The Tree of Life and the Tree of Being stand for many things: purity, justice, immortality, judgment, Divine design, and something that is seemingly unobtainable and even untouchable for humans. These Trees are a living testament to God's being, and thus they are comprehensible symbols of concepts that we cannot grasp. It is not a coincidence that, in both Christian and Muslim tradition, the Throne of God is entwined in the branches of the Tree.[9] God has located Godself in the arms of a very earthly Creation, whose true value and meaning are only revealed to those who believe. By believing, one can thus see the logic and balance of Creation itself. For this reason, the Trees are in fact a metaphor for the holiness of all Creation and the wisdom if its Author. At the same time the Trees are a sign of judgment upon those who are unable to grasp this truth.

Pablo Richard reminds us of the simultaneously liminal and fixed nature of the Tree of Life and the city in which it sits in the center. He writes:

> There is no end of the world, but rather a new creation. The new heaven and earth and the new Jerusalem is the final stage of history, the triumph or

complete fulfillment of history. It is a transcendent state: it lies beyond any human feasibility and beyond death, and yet it is part of our one sole history.[10]

It is in the very center of this new city—this new way of living—that the Tree of Life is rooted. Its branches and the fruit they hold stand as an invitation to eat and to celebrate the extraordinary gifts God has offered to all Creation. In the same way the Tree of Being is an invitation to know God and God's creation more fully, for the two cannot be separated. Ibn al-'Arabi reminds us of this fact when he writes:

> ... the density of the lower world [Earth] is as the density of [Adam's] corporeality, and the refined nature of the upper world [the heavens] is as the refined nature of [Adam's] spirituality. Thus on earth there is no mountain among those set on the earth to be stabilizers, but is in the position of [Adam's] bones which were appointed to be the stabilizers of [Adam's] body, nor is there therein any swollen sea, continuous or discontinuous, sweet-watered or not, but in the position of the blood in [Adam's] body, running in the runnels of [Adam's] veins, or standing in the channels of [Adam's] limbs.[11]

Thus the balance, cohesion, and logic of Creation reflects directly its Divine source, as modeled in the body of Adam. It is therefore only logical that Creation must be treated with the same respect as is accorded to God. By the same logic, disrespect toward Creation is disrespect toward its Author. This is the design of the biosphere in which we all live.

3. *Creation was designed to support, feed, and heal all of its inhabitants. When Creation fails to do this, it is due to human selfishness, jealousy, and sinfulness*
From a worldly perspective, the two Trees of the Bible and the Qur'an represent a utopian ideal. They have the ability to feed all people, to heal the wounds of all nations, and to model the balance that is inherent to the design of Creation. In this respect, these Trees are simply metaphors for Creation in its entirety *as it was designed to function*. Therefore, the Tree of Life and the Tree of Being only appear extraordinary in the face of the tremendous imbalance that humans have perpetrated upon the biosphere. Ibn al-'Arabi underscores this notion when he writes:

> If, therefore, you look at how the shoots of the Tree of [Being] differ, and at the kinds of fruit it has, you will recognize that the root thereof springs up from the seed of *kun* [being], separating out of it.[12]
>
> Now the good fortune of every creature derives from that word *kun*, from what each knows of the form of its letters, and what [they see] of the secrets hidden within it.[13]

The Tree, therefore, offers more than simply survival; it has wisdom to be learned and a deep joy to experience that could only be possible in communion with God. The secrets of the Tree are many—just as numerous as our questions regarding how to survive both spiritually and physically. To examine the meaning of *kun* is thus to explore the very fabric of how all life came into being, and at the same time to meditate upon how all the creatures of Creation are to live together in cooperation rather than conflict.

These very same yearnings are also addressed in the testimony of the Biblical Tree of Life. While the Tree of Life in Genesis is forbidden to the humans, it returns for Christians in Revelation to point in the direction of humanity's future as God envisions it. In the words of Pablo Richard:

> It is a logic rather than a chronology [that] depicts the logic of of God's final intervention in history: it is eschato-logical rather than chrono-logical. It shows the *ultimate* direction of history, where we are headed, and how the present age comes to an end . . . The future is anticipated. Likewise, the revelation of the future of history is not intended to satisfy unhealthy curiosity, but to make us live in the present in a different way. The importance of having a utopia lies not in the utopia in itself, but in how this utopia guides our present history and practice.[14]

Seen in this light, the Trees must be seen as humanity's charge to enact substantive change in our views of ourselves, greater Creation, and God. Thus, the Trees implore us to reject our human tendency to question the possibility of change. The Trees set a new standard, and offer a roadmap in the balance they represent, which is intended to guide humanity to a new way of living. Therefore, we are being called to live within the fabric of Creation as contributors to its health, rather than as agents of its destruction. This is not simply a commentary on nonhuman Creation; it is a specific claim that humans can live in balance with each other and with Creation as a whole. It is inherent in our very design. We are only prevented from such a state of being by our greed, our egos, and our narcissism. As Richard would say, the example of the Trees testifies that these inadequacies can be overcome through faith and action in our "present practices."

4. Two of the most followed figures in human history, Muhammad and Jesus, emerged from a Creation that they were called by God to exegete for their adherents

One cannot speak of the two Trees without speaking of the founders of Christianity and Islam; Jesus and Muhammad. The stories of these two figures are tied together with the Trees. Ibn al-'Arabi uses his writings on the Tree of Being as a platform to discuss the nature of the Prophet. So too, one cannot speak of the stories of Revelation and the Tree of Life and omit the image of the lamb (Jesus) who guides all believers to the Throne of God entwined in the branches of the Tree. By Muslim tradition, this is also the role of Muhammad, who will in the final days guide the believers into heaven.

However, before we can consider these eschatological stories, we must first admit that the very earthly lives that were led by Muhammad and Jesus were, in and of themselves, explications of the meaning and form of Creation. Many of Jesus' most famous parables used a farmer's understanding of the Earth to explicate the will and nature of God. Likewise, nearly any Muslim will tell you that Creation itself is full of signs or *ayats*, the very same word that is used to describe each verse of the Qur'an. This method of teaching, employed by both the Christian and Muslim sacred texts, points to an observation made in the previous chapter: we must view all of Creation as a Divine text, in the same way that we look to the lives of Jesus and Muhammad to tell us how to live our lives in a manner that honors Creation and its Creator. In this way, acknowledging the balance that is inherent to Creation and acting on such knowledge is a very tangible act of reverence.

In the Christian tradition, the Tree of Life ultimately comes to grow in the center of the New Jerusalem. Rather than a Temple, and rather than some other human-fashioned structure, it is the Tree that anchors God's city. From beneath this Tree flows water that can quench the thirst of all people, especially those who were denied access to water during their earthly lives. As Pablo Richard has so eloquently put it:

> Water here symbolizes the life that God offers to all who seek it. The interesting thing is that it is offered as a gift, that is, money has nothing to do with it.[15]

Recognizing Creation's balance clearly requires a new understanding of what is and is not valuable. This is one of the grounding claims of Ecological Realism. The gifts of Creation were designed for all its inhabitants regardless of their financial status. God made such an invitation through the lives of Jesus and Muhammad. To accept such an invitation is therefore to struggle to embody the balance the lesson of the Trees strives to convey.

Muslim tradition fortifies Christian tradition on this account. As we have seen, the body of Adam, which teaches Muslims so much about the nature of God, is reflected in the very Earth itself. Thus Creation is imbued with signs that point to the way we are all called to live: this is the revelation of the Qur'an and this is the message of Hadith. We are all intrinsic members of a Divinely fashioned Creation, which we were never given permission to destroy. Our actions with regard to the Earth thus reflect our level of respect for God and the complexity of Divine Creation. For many Muslims, there is no better testimony to the existence of God than the intricacy and symmetry of earthly Creation and the universe as a whole. As Ibn al-'Arabi has written:

> Then a wall was set around the tree, limits were fixed for it and it was given shape. Its limits are the [six] directions, viz. up and down, right and left,

before and behind, so that what is highest is its upper limit, and what is lowest is its lowest limit. As for the things that give it shape they are the horizons, the celestial bodies, the terrestrial lands, the laws [by which things are governed], the marks [by which they are recognized], the names [by which they are known]. So the seven zones were set to be as it were leaves giving shade, and the stars set in their courses to be as it were flowers in the celestial horizons, while night and day were set as two changing curtains, one of them black so as to veil things from sight when drawn, and the other white to be drawn so as to reveal things which may be seen. The Divine Throne was set as a treasure house for this tree and as a store house for its arms, a house from which is to be sought the succor of all the goodness it contains. In it are the attendants and the servants of this tree.[16]

Thus we see a Muslim vision of Creation that reflects its Author, and has entertwined within it the reality of God. Creation itself is seen as a text that proves not only God's existence, but also the cooperative manner in which Creation's inhabitants are designed to live. God's inherent balance is thus translated to Creation as a lesson for all to see, reflect and act upon.

Points of Disagreement: The Lessons of the Tree of Life and the Tree of Being (Shajarat al-Kawn) in the Biblical and Qur'anic Accounts

1. *While the Biblical Tree of Life is planted in the Earth and grows upward, the Muslim Tree of Being is planted in the celestial Heavens and grows downward to Earth, where it is met by its infernal counterpart*

As the English interpreter of Ibn al-'Arabi's *Shajarat al-Kawn*, A. Jeffries notes that Islam is often given to depicting a close parallelism between heaven and hell. As Jeffries notes in the introduction to al-'Arabi's text:

> The [Tree of Being] is often pictured as an *arbor inversa* whose branches reach downwards through all the circles of Paradise, and in this passage Ibn al-'Arabi is suggesting that there is so close a relationship between Paradise and Hell that there is an intertwining of their various elements.[17]

Hence, the Tree of Being, rather than being set apart as something that cannot be touched by those labeled "sinners," is in fact quite accessible to all, while also being entwined in the branches of its diabolic counterpart. As the Tree itself grows down from the heavens to Earth, one might therefore presume that its trunk and roots are untouchable to any mortal. It is the inverted "top" of the tree that is within the reach of Earth's inhabitants. In contrast to the Biblical Tree, the fruits of the Tree of Being contain both rewards to the followers of God and punishment to sinners. As the Tree of Being's branches comingle with its infernal counterpart, the space that is created in the two Trees' meeting is what one could call a Muslim *saeculum*. In other words, the place where humans are obliged to live out their earthly lives: a place of liminal and temporary location within the intersection of heaven and hell.

The Christian Tree of Life points to another cosmology entirely, one in which there is a clear separation—and distance placed in between—the sacred and the profane. As we saw in the Biblical passage, the Tree of Life in Genesis is ultimately guarded by a cherubim and a flaming sword, in order to ensure that the humans whom God ejected could not touch it and gain eternal life. So too does the Tree of Life remain out of reach throughout the Christian Bible, until those chosen to enter heaven are allowed to eat from its branches. Unlike the Tree of Being, the fruits of the Tree of Life are reserved exclusively for the people of God. Being excluded from these fruits is seen as punishment. Thus, the Tree of Life initially appears on Earth as something that cannot be defiled by earthly (sinful) human touch, a characteristic echoed in its ultimate eschatological presence. For this reason, one could say that the Christian Tree of Life is always growing upward, be it on Earth or in heaven, and God alone determines who is rewarded with the right to approach it. Like the Tree of Being, the Tree of Life can easily be interpreted as a vehicle for the realization of God's justice. Yet in the Christian version of the Tree, it is implied that Divine justice is something that is meted out in a new context, on a new Earth in God's own city at the end of human history.

2. *In Islam there is a distance and yet a profound nearness of God to humans and to Creation. For Muslims, God could never become an earthly corporeal being as Jesus became for Christians. Thus the signs of God on Earth arguably take on a different form for Muslims, as God's will, desire, and designs are manifested through God's earthly Creation, God's revelation of the Qur'an, and the life and teachings of God's final Prophet, Muhammad*

"How are we to live?" is a question asked by both Muslims and Christians in light of the teachings of their respective traditions. For the Christian, one must first look at the earthly life, death, and resurrection of Jesus as the story that binds all Christians together, regardless of their individual traditions. Seen as a manifestation of God living upon Earth in human form, it is understandable that Christians focus their attention on the teachings of Jesus and the subsequent writings of his followers in order to draw out life lessons. Depending on one's own christology, Jesus can be seen in a variety of ways. These views range from a very earthly, poor, and marginalized teacher of truths to an almost otherworldly figure, triumphant over human laws and even death itself. Whatever their christology, Christians claim Jesus, as God living among humans upon Earth, the Messiah who was ultimately killed for his teachings and actions because they were seen as a threat to those in power and the status quo in general. In this way, Christians claim for themselves a very earthly manifestation of God—one who took on human form and suffered as humans do. It is thus through this hermeneutic of Christ that many who call themselves Christians can most easily see God as being connected to the Earth.

For the Muslim, such an image of God—as one who lives as a human and even dies—is blasphemous. The God of the Muslims could never be killed, nor be subjected to the indignities of Jesus. Such a notion is an irredeemable insult to the Muslim understanding of God. To Christians, the Muslim understanding of God might appear distant, or even cold and judgmental. Yet this is not at all how a Muslim views his or her Creator. For the Muslim, God becomes intimate through the reading of the revelation of the Qur'an, which is widely understood as God's direct and unimpeachable Word. The love and mercy of God is also reflected through the *umma*, or Muslim worldwide community. At the same time, God's nearness is expressed through Creation itself, in the intricacy, beauty, and inherent balance of its design. As great as he was, Muhammad is still very much a man for the followers of Islam, though the most extraordinary of men.

Thus it is through differing, contradictory understandings of God and God's manifestation on Earth that Muslims and Christians are drawn to seeing themselves as part of a Divine Creation. While the Muslim God would not live as a human upon the Earth, the Muslim God is infused in every fiber of Creation. In the absence of an earthly Christ, one might argue that the Muslims are drawn to look even more closely at Creation in order to learn its lessons. At the same time, Christians cannot abandon their own responsibility to the Earth as if such an obligation was embodied in the earthly life of the God who lived among them. Jesus' life on Earth draws attention to the importance of life—all life—on Earth and humanity's responsibility to celebrate and protect it. Thus the Muslim and the Christian are called to the same set of responsibilities, but through different Divine means. Whether speaking of the Qur'an or the Bible, Jesus or Muhammad, one is confronted by a God of Creation who imbues the human and nonhuman Creation with the breath of life and the possibility of dynamic change.

The Ecological Footprints of Spain and Morocco: The Seas

When Wackernagel and Rees first envisioned the project of the Ecological Footprint, they did not believe that it would be possible to quantify the consumption levels of a country with regard to the ocean's resources. So many factors remained unknown: the exact capacity of the seas on a yearly basis, the exact tonnage of catch by a nation-state, and the hectares per capita of each citizen, given that territorial waters are often contested. Since Wackernagel and Rees' original work, however, some have attempted to answer these questions. Chief among them is the organization Redefining Progress, which works in cooperation with the World Wildlife Fund. The following statistics represent a new evolution in the calculation and use of the Ecological Footprint. The new calculations include the *yield factor*, in which a specific country's production in a particular area (i.e. cropland, forest, ocean) is compared to the average production for the entire planet. Also introduced is the *equivalency factor*, which places the ratio of the average production of a particular nation's land or ocean beside the average productivity of the world's lands and oceans. These new sets of statistics render the *bio-capacity* of particular areas within a nation—what a particular type of land or ocean can sustainably yield without being damaged or permanently depleted.

Spain

The consumption of fish in Spain is among the highest within the EU. According to an Ecological Footprint compiled by the Redefining Progress and the World Wildlife Fund, the last recorded annual marine fish and seafood catch totaled an extraordinary 1,399,699 tons of sea life.[18] When this is added to Spain's freshwater catch, the total fish production of Spain totals 1,462,435 tons of fish.[19] When expressed in terms of an Ecological Footprint, the average Spaniard requires 0.56 ha/cap of ocean to meet his or her consumption needs.[20] This lies in stark contrast to what is now estimated to be the biological capacity of the territorial waters of Spain, which totals a decidedly small .04 ha/cap.[21] Spain's annual catch, including exports and domestic consumption totals .36 ha/cap.[22] The territorial waters of Spain sitting on a continental shelf (the richest fishing grounds) total 6,210,000 ha, or .17 ha/cap.[23] Despite the tremendous amount of ocean life Spain extracts from the sea on an annual basis, Spain is still a net importer of fish.[24]

Morocco

The consumption of fish in Morocco is much lower than in Spain. According to an Ecological Footprint compiled by the organization Redefining Progress and the World Wildlife Fund, the last recorded annual marine fish and seafood catch for Morocco totaled 220,629 tons.[25] When added to Morocco's freshwater catch, the total fish production of Morocco totals a relatively modest 222,590 tons of fish.[26] When expressed in terms of an Ecological Footprint, the average Moroccan requires .09 ha/cap of ocean to meet his or her consumption needs.[27] This level of consumption stands far below the estimated biological capacity of the territorial waters of Morocco, which is .29 ha/cap. Morocco's annual domestic catch, including imports, exports, and domestic consumption totals .12 ha/cap.[28] The territorial waters of Morocco, which sit on a continental shelf, total 7,040,000 ha, or .25 ha/cap, a decidedly larger area than Spain's.

From these statistics alone one can conclude a number of things. First, Spain's consumption levels of the ocean's resources dwarfs that of Morocco. Second, Spain's allotted area for domestic fishing is significantly smaller than Morocco's, and the bio-capacity of those waters are already tremendously depleted. In contrast, Moroccan consumption of ocean life is markedly below its capacity, while its allotted ocean area footprint is much larger per Moroccan citizen. Given these circumstances, one can understand Spain's interest in renewing the now defunct fishing treaty that it once held with Morocco. Moroccan waters have the potential to make up for Spain's depleted capacity and its large appetite for ocean life. Despite this attractive proposition, Spain's consumption of fish would be only temporarily sated if Spain were given free run of Moroccan waters. Spain's level of fishing and capacity for consumption might well deplete Moroccan waters as it has depleted its own. These numbers are more easily seen when expressed in a graph:

Spain—Harvest and Consumption Ocean Footprint (expressed in hectares per capita divided among the Spanish population)

Annual Consumption	Production (Catch)	Fishing Area	Biological Capacity
.56	.36	.17	.04

Morocco—Harvest and Consumption Ocean Footprint (expressed in hectares per capita divided among the Moroccan population)

Annual Consumption	Production (Catch)	Fishing Area	Biological Capacity
.09	.12	.25	.29

The Conflict Over Natural Resources: The Fish Wars

Touba Belkadi

In your religion and your tradition, is there any way to speak about the beauty and holiness of the land?

In our tradition we say a proverb: "he who doesn't save his land is as good as dead." And we have another proverb: "he who does not plow his land . . ." It's like that. The land in general, in our religion we say a verse from the Qur'an: "God created the skies and the land." So we are on the land. The human on land. And the best creature is the human, who was created from land. So he should be useful to the land, not just living without meaning.

Is there a way to understand the greatness of God in Creation?

Thanks be to God—His greatness is evident everywhere on Earth. But it is most evident in the brain that He gave to humans. The best thing Allah gave to the humans is the brain, which is used to think about some things which are extraordinary. And you ask yourself, how does the human do that? God gave the human that brain through which he discovered how to make rain fall using science. What can you say? The brain is the best thing God has given us.

How do people speak about the blessings that come to us from the land?

God created the land, the mountains, and the seas. The greatness shows in the seas, which provide us with the fish that we eat. And also the land as well, from which comes all that we eat. From Earth comes water. All these riches for the human. The land is everything.

How do you look at Spanish people?

I like them. Why do I like them? Because they look like Arabs, like us. I find them to be like us — beautiful and handsome. And they are clean. They speak like us — fluently. And they say that there are the Andalucian roots. And the things they have in Spain are similar to what we have. They like what we like. There are some people who are generous, like Arabs. They have the same education. We like them, although they have taken our land and our fish.

Touba Belkadi, 32, P. E. teacher, Fes[29]

Mercedes Carretero Ortego

What do you see as your connection, if any, to the people living in Morocco?

Well, I think that with the Andalucians they have some things in common, such as the character, their way of living their life, it doesn't matter what you have, but [that you] enjoy life. Their landscapes, well, their villages are very similar to the ones in Andalucia, the houses, the color, the light and even, in many cases, the physical characteristics. For example, the dark hair, their expressive eyes, their skin, although it is a little bit darker.

What are your biggest differences?

Maybe they are less advanced than we are, what I know about Morocco at least, they live in the past centuries, there are also economic differences, different infrastructure too: villages, roads, medical differences, labor differences . . . I think that they are further behind, less developed.

You have already been in Morocco, haven't you?

Yes, I've been there once.

Would you like to go back to Morocco?

I would love to.

Why?

I like its white towns [an allusion to the fact that many Moroccan villages resemble the "pueblos blancos" or "white villages" of southern Andalucia], *the quietness of the people when they talk and enjoy their day; they are all in their place, they are not in a hurry, but never stopping. I like the fact that they have time to dedicate to other people: talking, having a drink . . . I like their buildings, their neighborhoods, their mosques . . . I like their style.*

What do you think would be the most important thing to do to improve Morocco's relationship with Spain? Imagine that you are the king of Morocco or José Maria Aznar and that you can change whatever you want in order to make these connections better.

I have no idea about this. To improve the relationship? I don't know, perhaps I would establish an exchange, I can give you what I have that you need in exchange for what I need that you have, an exchange, cooperation on both parts. I think that they have some economic needs, and Spain and other countries could help them. They also have the sea we want for our fishermen to fish, that fishing zone. Another thing would be to open the frontiers, not to keep them in there as they are but give them the chance of moving, . . . let those who want to leave in search of something better, who want to change, to look for a way out to other country. At the same time we can learn many things from them, their way of thinking, their kindness to others . . . I don't know, an exchange program between countries . . . but maybe there are many political interests behind [preventing] this.

Mercedes Carretero Ortego, 36, Spanish teacher, Seville[30]

The topic of the so-called "fish wars" between Morocco and Spain has a long and lively history of its own; one in which the struggle for control of diminishing natural resources is overtly played out. According to the political theorist Richard Gillespie, the Moroccan–Spanish conflict over fishing grounds has long "served as an approximate barometer of the state of relations between the two countries."[31]

The relationship Morocco and Spain forged over fishing rights has, since Moroccan independence, been focused on Spanish access to the rich fishing grounds off Morocco's extensive and biologically rich coast. Spain's own territorial waters have long lost their capacity to supply Spain's marine consumption levels. Regretfully, at no point in Spanish–Moroccan relations has this issue been resolved to both sides' satisfaction. Beginning in 1957 an agreement—drafted but never ratified—guaranteed both sides access to the other's waters.[32] In 1962, Morocco unilaterally extended its territorial waters from 6 to 12 miles, a choice that Spain refused to respect. This set the tone for a number of ensuing conflicts,[33] which have included the seizing of Spanish boats by the Moroccan government, a levying of fines for what the Moroccans termed illegal fishing, and protests by Spanish fishermen in southern Spain.[34]

Spain and Morocco's fishing disputes have long been linked to land disputes. For example, the 1969–1970 Spanish withdrawal from the Moroccan city of Ifni resulted in Morocco renewing Spain's right to fish off their Atlantic coast.[35] Unfortunately, Spanish fishermen persisted in using illegal nets and often refused to recognize the limits Morocco placed on the location of their fishing or the size of the catches taken.[36] After the 1975 Green March when Morocco reclaimed the Spanish Sahara (now known as the Western Sahara) from Spain, Spain chose to look the other way in exchange for an extension of Spain's fishing rights along the coast of the disputed territories.[37] This decision was a great disappointment to many Moroccan fishermen, who felt that their own government had compromised their industry in the name of resolving land-based territorial claims.

Spanish anxiety about the possibility of conflict regarding the Spanish possession of the Two Cities guided Spain's fishing negotiations with Morocco. As a result, more than one Spanish government has soft-pedaled its disagreements with Morocco over fishing and Morocco's continual modification of its demands on its northern neighbor.[38] Thus the unilateral extension of Moroccan offshore fishing limits were met with few outward protests by Madrid.[39] This was a difficult and potentially expensive reticence on Spain's part, because the votes of southern fishermen were needed to preserve a majority in the Cortes.[40] By the time of the Socialist ascension to power, the fishing disputes had taken on new dimensions, including Morocco's attempt to negotiate a linkage between fishing rights and the free transport of Moroccan citrus fruits across Spain.[41] Meanwhile, off the increasingly disputed Western Sahara, Spanish boats were being attacked and Spanish fishermen were killed or kidnapped by the Polisario—the movement calling for Saharan independence from Morocco.[42]

By 1986, Spain had become a member of the EC and all subsequent negotiations regarding fishing were handled by Brussels, though Spain remained the greatest beneficiary of agreements, signed in 1988, 1992, and 1995.[43] Through these negotiations, Morocco moved toward obtaining tacit European approval of its occupation of the Western Sahara, by Brussels' acknowledgment of Spain's right to negotiate the fishing rights off the coast of this disputed territory.[44] However, these signed agreements did not end all disputes. Spanish boats continued to be caught taking more fish than the agreements permitted.[45] For these and other reasons, the 1995 treaty was never renewed, and the absence of any agreement permitting Spanish boats to fish in Moroccan waters is currently a cause of great friction between the two nation-states. Today, relations between Morocco and Spain have deteriorated in part due to the ongoing fishing dispute. Further trade relations initiated by Morocco, particularly in the area of agriculture, have been put on the backburner by Madrid because of the lack of a fishing treaty.[46]

There are also other disagreements that are pulling Spain and Morocco apart. Ongoing disputes regarding Spain's possible support of the Polisario Movement fighting for the independence of the Western Sahara, Spain's recent insistence on exploring for oil in waters claimed by Morocco between the Canary Islands and Western Sahara, the issue of the Two Cities, and the issue of illegal immigration between Morocco and Spain are all undermining Spanish–Moroccan relations. These combined points of dispute led the Moroccan government to recall its ambassador from Madrid in October 2001.[47] As of this writing, he has yet to return.

Through the rubric of Ecological Realism, one can clearly see that all the points of conflict between Morocco and Spain are linked by disputed claims to land and natural resources. Whether the conflict is over fish, the right to trade agricultural produce, control over the phosphates-rich Western Sahara, contested attempts to explore for oil, or the right of human mobility deeply linked to agricultural harvest, the diplomatic dispute being waged between Morocco and Spain has a common ecological basis.

Sustainable Diplomacy requires negotiations that embrace and engage rather than run from ecological conflict. By widening the vocabulary of the participants to include religious principles informed by ecological realities, a new tool for communication could be applied to what appear to be intractable differences. At the same time, Ecological Realism asserts that Northern methods of globalization are not necessarily the means by which conflict can be resolved; in fact, such measures are often only useful in maintaining an uneven playing field across the North–South split. Thus, simple applications of trade enhancement on Northern terms is not a probable or just solution for the Moroccans, and neither will such tactics provide the firm foundation Spain (or the EU) will require for building a long-lasting peace across the Strait of Gibraltar. While a just resolution of the Two Cities dispute could open the door to a series of new and potentially productive dialogues between Spain and Morocco it is, in and of itself, not enough. For this reason, creatively and directly addressing the fishing dispute could provide a means and a context for further negotiations on other points of resource-based disputes, and could establish a framework for resolving conflicts over oil, phosphates, land, and human mobility.

With this goal in mind, let us turn to the question of fishing rights, as we engage the framework established by our examination of the Two Trees.

Common Ethical Principles: The Lessons of the Tree of Life and the Tree of Being for Spanish and Moroccan Relations

1. We live precariously and temporarily between heaven and hell, both of which can be reflected in earthly life depending on the path of human action

The fishing dispute between Morocco and Spain is a pivotal conflict. The terms of engagement established in this disagreement have informed the unsuccessful negotiations between Morocco and Spain in all of their other points of conflict. Thus, like the question of the Two Cities, resolution of the fish wars is a key to building a deeper level of respect and understanding between Morocco and Spain and their respective populations. The dimensions of the conflict are profound in how they affect the lives of many ordinary people, including everything from their livelihoods to what they can find to eat in their local markets. It is for this reason that most Moroccans and Spaniards have an opinion on this subject. The wealth of phosphates and oil are out of the hands of the average citizen. Fish, on the contrary, are a very tangible asset.

The precariousness of the relationship between Spain and Morocco cannot be overestimated. The choices made in the next few years will reverberate between the two respective populations for years to come. Resolution of this problem must therefore be a top priority among those who represent each nation, as well as third parties who might be able to provide a new vocabulary and context for discussions. The Ecological Footprint may be one place to begin to call for all countries to scrutinize the dimensions of their own ecological interdependency and their potential theft of other nations' sustainability. As we have seen, it is Spain's large appetite for the products of the sea that has helped to set the stage for this conflict. The average Spaniard's consumption of seafood outstrips the average Moroccan's by approximately six times and so it is no wonder that Spain has reached beyond its own territorial waters to meet the lucrative market demand of its population. Were Spain limited to its own territorial waters, the average Spaniard would have to limit his or her annual purchase of sea products to one-fourteenth of current levels of consumption.[48]

With many of its boats sidelined due to ongoing disputes, the price of fish has risen in Spain and will continue to do so as decreasing supplies meet increasing demands. These disputes have led Spain to seek new fishing grounds off the coasts of other African neighbors, including Mauritania and Senegal. Already thought to be over-fished, the continental shelf of Mauritania was leased in 2001 to the EU in a five-year agreement principally benefiting Spanish fishermen.[49] At the same time, Spain's EU membership status has given it the right to fish off the coasts of England and Ireland, both of whom had moved to tighten EU fishing regulations in anticipation of Spanish membership.[50] Still, the agreements signed have not been able to supply enough fish for Spain, who has gone so far as to strike an agreement with Chile in order to secure a long-term source of fish.[51]

Spain has clearly established an untenable pattern. Spain will ultimately be obliged to live more closely within its Fair Earthshare regarding the sea or it will be a major collaborator in the exhaustion of global fishing grounds. Spain's current approach to obtaining the rights to fish is clearly an expression of ecological colonialism whose goals are short-term gain for Spain at the expense of any who allow Spain to harvest from their territorial waters.

Morocco's own efforts to exploit its newly reclaimed ocean territories are also not without problems. Plans are now being made to significantly expand the Moroccan fishing industry, including the construction of six new "fishing villages" in the Western Sahara. Through the expansion of its fleet and the building of new refrigeration units, satellite tracking, upgraded boats and ports, the Moroccans have aimed to become one of the "top 15 fishing nations" in the world by 2003.[52] While one cannot blame the Moroccans for wanting to profit in an area that had until recently been off-limits to them, their probable duplication of European practices will ultimately result in the same level of ocean exhaustion. Thus, establishing a new approach to fishing, as well as new strategies for bioregional cooperation and mutual respect must be found if Morocco and

Spain are to find a way to ecologically coexist. This is the work of Sustainable Diplomacy: finding a path whereby domination can give way to workable, realistic ecological cooperation.

2. *There is an* axis mundi *in God's Creation, providing balance, wisdom, cohesion, and logic to the biosphere. It can be seen and understood by those who look for it and strive to interpret it*

The balance of Creation reveals the depth of human and nonhuman interdependency. The current fishing policies of Spain, and probable future approach of Morocco are dictated by a market economy in which long-term value is hardly considered. As the lessons of Wackernagel and Rees have shown, an economy based on human-made currency structures rarely reflects the value or capacity of Earth's actual resources. The imbalance of current global fishing approaches are clear to all who examine them. The means by which we all, including Morocco and Spain, can play a role in breaking long-term patterns is harder to comprehend. Admitting the reality of ecological imbalance is the beginning of building a new type of eco-centric power. Knowledge of how to bring balance is the locus of true power, one that surpasses all others.

The true *axis mundi* of the Earth has been replaced by the *axis mundi* of consumption and short-term greed. The balance of Creation rejects the priorities of a globalized economy, which sees no limits to the transportation of goods and services between their origin and their market. Yet we are blind to these truths as long as we accept the individual nation-state as being the final arbiter of consumptive power. Bioregions are a truer lens for viewing lands, seas, and peoples and for conducting diplomacy. Regardless of the existence of the EU, Morocco, and Spain are living in a common bioregion. They share the same fishing grounds, just as their people share a common political, economic, religious, and genetic history. Thus, balance cannot be legislated so much as it can be recognized and followed. The Moroccan and Spanish populations together must, as Larry Rasmussen has written, "turn to Earth" for their guidance, or risk perishing at the hands of short-sighted, individualistic national policies. This change in behavior will require a new level of commitment and a new depth of ecological literacy.

3. *Creation was designed to support, feed, and heal all its inhabitants. When Creation fails to do this, it is due to human selfishness, jealousy, and sinfulness*

How could the fish wars become a turning point in Spanish–Moroccan relations rather than repeating patterns of noncooperation? What barriers must be crossed, and what patterns must be broken? How could resolution of the fish wars serve as a template for resolving other conflicts between Morocco and Spain with regard to the control and distribution of natural resources?

Thus far, negotiations to reach a settlement that would allow EU fishermen (principally Spaniards) to work in Moroccan waters have not realized a satisfactory conclusion for both sides. However, within the negotiations that have taken place, seeds have been planted for a sustainable solution that honors both the reservations of the Moroccans and the European desire to fish close to its landmass. Although the following terms proposed by Morocco and rejected by the EU have not ended the impasse, they merit examination due to their acknowledgment of the limited nature of the ocean's resources. The proposals put forward by the Moroccans included:

1. a reduction by one-half of the European fleet in Moroccan waters; 2. European vessels to be trackable by satellite; 3. a sharp reduction of total allowable catches for the European fleet—particularly for high-value cephalopods (octopus and squid); 4. all catches by the European fleet to be landed in Moroccan ports; 5. 25% of the manpower on European vessels to be Moroccan; 6. the accord duration to be two years—compared with four years for the old agreement; and 7. payment of $125 million US per year for fishing rights—the same as before despite the reduced quotas.[53]

Given that Spaniards are consuming 14 times the amount of sea life that their own territorial waters can sustain, reductions far beyond those proposed by Morocco are reasonable. Yet within Morocco's proposals stand a number of features worth retaining: drastically reducing consumption levels, mutual accountability, transparency, acknowledgment that ocean life is limited, acknowledgment that both Spain and Morocco are dependent on waters within the same ecosystem, and the notion that Spanish and Moroccan workers should ply their trade side by side.

The building blocks of Ecological Realism demand the construction and acceptance of treaties, which reflect the facts of the biosphere as it actually exists. Therefore, any treaty that reflects the goals of Sustainable Diplomacy will take its cues from the design of Creation. Such a fishing treaty between Spain and Morocco will include incorporating "rest periods" for the seas to recover from human harvests, gaining a new understanding of the advantages of sharing and maintaining natural resources as a bioregion, emphasis on the importance of retaining healthy fish stocks in close reach of both countries, and building a mutual trust through cross monitoring and third-party monitoring of both nations' fishing practices. On the land, practices will have to change as well. Spaniards will no longer be able to consume fish at the rate to which they have become accustomed. At the same time, Moroccans will also have to be mindful of the biological capacity of their own waters and will have to modify their current plans to capitalize on their potential to become large-scale exporters of fish. The globalization of the Moroccan fishing industry is not sustainable over the long run. The Ecological Footprint proves this to be true.

4. *Two of the most followed figures in human history, Muhammad and Jesus, emerged from a Creation that they were called by God to exegete for their adherents*

A new Moroccan willingness to negotiate with Spain could result in larger and more reliable markets for Moroccan agricultural products. While such an exchange is attractive on a number of levels, a simple quid pro quo exchange cannot be the goal of Sustainable Diplomacy. Instead, the design of ecologically realistic exchanges and cooperation between Spain and Morocco must be grounded in the goal of establishing and maintaining a locally grown and locally consumed philosophy of consumption.

It is unsustainable for Spain to travel to Chile or Canada to find fish, and for Morocco to rely on the United States for the grain it cannot produce itself. The Spanish and Moroccan goals of singular self-sufficiency must be translated into an ecologically more achievable mutual self-sufficiency, even as Spain finds itself more and more integrated into the EU. For this reason, the Moroccan–Spanish relationship must thrive in order to ensure the future health of European–Maghrebi relations. At the same time, the Moroccan–Spanish relationship must survive and flourish in order to play a role in repairing the damage and preserving the beauty of the results of each nation's imposition into the affairs of the other. This mutual intercourse over many centuries may well have left Spain and Morocco in the best position to teach their neighbors how to live together.

Jesus and Muhammad not only exegeted the land, but they also exegeted the people who inhabited it. They subsequently told truths both about the people, the land, and themselves in the course of their lives. Muhammad and Jesus also both taught that all humans were created to live together under the leadership of one Creator God. In many cases they taught that this Divine leadership was most tangibly expressed in Creation itself. They taught that through the seasons (each a time of anticipation), birth, growth, harvest, rest and death, the lessons of how Creation functioned are conveyed. Through the fruit of trees and through the fish of the seas these lessons are also reflected. While Jesus reminded his followers of an intimate God who knew the number of hairs on each of their heads, Muhammad taught of a generous God who was willing to share all Divine knowledge with any human who was willing to listen, learn, and apply such lessons. The balance that is Creation is reflected in Muhammad's and Jesus' teachings; yesterday, today, and tomorrow. One of our greatest problems is our growing ability and willingness to undermine this balance for short-term gain. In sharp contrast, Jesus and Muhammad taught lessons to be drawn upon for the long term, for a people who are daily tempted to do exactly the opposite.

Points of Disagreement: The Lessons of the Tree of Life and the Tree of Being for Moroccan–Spanish Relations

1. *While the Biblical Tree of Life is planted in the Earth and grows upward, the Muslim Tree of Being is planted in the celestial Heavens and grows downward to Earth, where it is met by its infernal counterpart*

The disparity in the accessibility and distribution of natural resource wealth between Spain and Morocco mimics the lessons of their respective

Trees. For the Spanish, the fruits of the Tree are accessible only to God's chosen people. For the Moroccans, the fruits are designed to be accessible to all, though through human choice the number who actually taste this fruit are radically pared down. Such a theological explanation goes some way toward explaining Spain's and Morocco's respective stances on the issue of fishing rights. For many years, the Moroccans, for a price, shared their bounty with the Spaniards. In turn, when the arrangements no longer suited Spain, the Spaniards went across the globe to maintain their unsustainable levels of consumption, as if it were their God-given right.

Now both sides must acknowledge that they live within the same saeculum: between a paradise of sharing and abundance for all, and the hell that will emerge when fishing stocks are depleted beyond repair. For some time now, it has been the Moroccans who have spoken more forcefully of conservation of fishing grounds, yet their new plans to develop their own fishing industry on a grand scale betray that commitment. While the Spanish through the EU press for a broader swath of ocean territories to fish, the time has now come to return to a principle that both Moroccan and Spanish religious cultures claim to be true: that long-term rewards are to be gained by those who can distinguish between honoring Creation and contributing to its destruction. For this to happen, the Spanish—and with them the entire EU—must let go of over a millennium of believing that they are the chosen people, who have the right to make choices for everyone else. At the same time, the Moroccans must stifle their own temptation to mimic the actions of those whom they have so long criticized.

The upright Tree of Life and the *arbor inversa* of the Tree of Being symbolize the dichotomy of worldviews between the Spaniards and the Moroccans. For the Spaniards, the Tree grows upright as it should, representing an upright people whose cause is in conformity with Creation itself. For the Moroccans, the *arbor inversa* can be seen as reflecting a world that has been turned upside down since 1492, when the grandest chapter in the history of Islam was brought to a close, as European hegemony soared. The continued shame of European colonialism, be it traditional or ecological, reminds Moroccans of the turn that history took when they were forced to flee the European landmass. The time has come for this shame to end. The means to bring about its end lies in Ecological Realism's new way of understanding power, responsibility, and fruitful, worthwhile cooperation. For all the disagreement over ownership, access, and rights, there remains much agreement regarding the source of Creation itself, its fragility, and the necessity of our collective survival.

2. *In Islam there is a distance and yet a profound nearness of God to humans and to Creation. For Muslims, God could never become an earthly corporeal being as Jesus became for Christians*

For many Muslims, Christians are not monotheists. The Christian explanation of a triune God seems to be a new form of polytheism. While

Christians and Jews nonetheless remain "People of the Book" in Muslim eyes, the logic of the Christians' inclusion in this Book is based more on shared prophets and stories than on a unifying image of one singular God. As it has been noted, the notion of a God who could live on Earth as a human—and even be killed—is anathema to the Muslim understanding of God.

For many Christians, God's earthly embodiment in Jesus is one of the clearest signs of God's existence and corporeal reality. A God who cared enough to come and live among ordinary people as a poor man remains the Christian's benchmark connection between God and Creation. In this Divine embodiment, Creation was physically touched once again by a radically loving and defiantly present God. For most Christians, Jesus' life and death is *the* distinguishing characteristic of Christian belief. Without it, understanding of God's true nature is made less tangible and intelligible.

For both Muslims and Christians, God's nearness to Creation is an accepted fact. For both traditions, Creation is a Divine gift, and all of humankind is charged with helping to ensure its survival. The Muslims' lack of an earthly Messiah has not prevented Islam from presenting an embodied theology for its believers. Creation itself is the text to be read alongside the revelation of the Qur'an. So it is for Christians too, as Jesus's lesson from the parable of seeds falling on different soils is remembered and repeated, an earthly Christian faith challenges modern efforts to detach human life from the rest of Creation.

Despite any and all debates, an Earth-centered ethic is embedded in the text and traditions of Christianity and Islam. It is the work of Sustainable Diplomacy to lift up such interpretations as points of inspiration in constructing a new way of living together. This is why having knowledge of the religious traditions of one's conversation partner is so critical. Without this knowledge, looking at the land together can often become only a purely economic or political exercise. Ecological Realism requires a deeper level of conversation—one that brings religion, land, and life as it is led by the ordinary people into the center of discussion. Thus, Sustainable Diplomacy is impossible without a new set of priorities, abilities, and commitments.

CHAPTER 5

THE CONFLICT OVER PEOPLE: THE STORY OF ABRAHAM AND IBRAHIM AND THE STRANGERS, THE CONSUMPTION OF ILLEGAL HUMAN LABOR AND THE CONFLICT OVER IMMIGRATION

Sustainable Diplomacy demands that people become willing to abandon old perceptions and means of communication. While Ecological Realism underscores the common earthly connections that all creatures share, it also reminds us of the reality of one common human fate: ultimately, human populations will learn to cooperate and share natural resources in a sustainable manner, or perish.

Commonly held perceptions that must change in order to insure the biosphere's survival. Existing definitions of development, wealth, time, responsibility, language, and ancestry are only a few. The practitioners of Sustainable Diplomacy are therefore charged with finding words, definitions, images, beliefs, and ideas that can translate across existing borders and can make bioregions a politically and spiritually recognized reality. This change cannot be dictated from "above." It must come from "below," from the very populations whom leaders claim to represent. Sustainable Diplomacy is a systemic diplomacy, which becomes more and more adept at diminishing the importance of existing political borders. Sustainable diplomats will work to construct a dialogue between the populations of two or more countries, while their traditional counterparts concentrate on bettering communication and cooperation among heads of state. Through these parallel efforts, lasting relationships can be established and maintained.

Understanding the Conflict Over Immigration: Carmen Barragán Diaz and Jamal Rhmiro

Both of the following interviewees are well-educated professionals. Carmen Barragán Diaz is a feminist lawyer working in a small pueblo outside of Seville whose clients are almost exclusively women. Carmen's views are more progressive than many of her fellow Spaniards, and it is for this very reason that she is represented here. In contrast to many other interviewees, Carmen expressed neither a disdain for North Africans nor a desire to devise a system to limit their presence in Spain. Her ability to see Spain's own past in the Morocco's current situation is exactly the type of thinking upon which Sustainable Diplomacy must be built. Carmen has made connections that echo the views of other progressive people from her country. In many other respects, including her professed atheism, Carmen is representative of many Spaniards of her generation.

Carmen Barragán Diaz

How do you feel connected to the land where you live?

I feel connected to my land by means of my friends, that's really important to me, by means of the common history that we have, of the way of being and living, mainly the sense of life, the concepts of space and time. This is what connects me to the land where I live and that I share. And, very important, the persons I share the land with.

Then you would say that your friends are part of that land?

Yes, of course, my friends make the land.

Do you own any land?

No, I don't.

What do you see as your role in working for a good future for the next generation?

Well, that is very presumptuous. I don't know whether I work for the future of a generation. I work for the present of this generation.

But, do you think that your job will influence this future?

Actually, if we improve the present conditions of life, then we are also improving the future. In my case, working with women, if I can make them to improve their lives, this will mean that future women will live better and have better living conditions. But I am happy with improving the present, because I think that it is to sow for the future. I do not work looking so far ahead; rather I work to improve this moment that we are living now. I believe that little by little we will climb the steps to reach the future.

Do you see their future as being connected to the health of the land?

. . . I think that the land is worse and worse. We are destroying it more and more, there are less and less natural areas, and I think that future of the next generations is quite dark. In this sense I'm really pessimistic. I think that we are not aware of what we are doing. We are destroying the planet.

What does your religion and/or cultural traditions teach you about the origin of the land/Creation?

If we talk about religion, I should say that I'm atheist although my education has been completely religious, I studied at a nuns' school where they ladened me with religious rites and beliefs. Now nothing is left, I've spent my entire life trying to erase that, all those beliefs that I do not believe right now. But they never disappear completely and you carry them around for the rest of your life and I would like to forget it completely.

If you had children, what would you teach them about the origin of the Earth?

I will never have children because I never wanted to have them, that's clear to me. If we talk about teaching something different to the next generations, I would try and teach them to be more understanding, to have more unity. I would teach them a wider sense of collectivity, less individualism, and less egoism, a good sense of tenderness, of feelings, and to be less materialistic and less consumer-oriented.

Now, I'd like to talk about your personal philosophy. What does your philosophy teach you about your connection to the land and other living creatures?

I think that we are all equal, that everything else [humans have created] *has been created and invented to separate people from each other. I really think that we are all the same, it doesn't matter if we are from Africa, Asia, or from any other continent. There are no big differences between us. We all think, feel, and believe the same . . . .*

Do you have special responsibilities regarding Creation because of your cultural tradition?

I don't feel responsible for Creation. I'm just here, I don't know the reason why I'm here, I suppose that it is because I come from nature, the same as plants or animals. I'm here and I'll try to do my best to conserve this world as well as possible and see-ing human beings going on as best they can. But I don't feel responsible, I'm not responsible for famine, wars, bad governments, dictators . . . I'm not responsible for that because I have no power, as I have no power I'm not responsible. In any case, I am a victim of that power, but never responsible.

In your philosophy, is there a particular way of speaking about the beauty and holiness of the land?

Beauty is something relative because everyone has his own sense of beauty. What is true beauty? Beauty may be either a feeling or a flower, which you may think that is beautiful, or a monument that you like. It is something very relative, I think that it is what we feel, it is a feeling, because many times we see a person who is not beau-tiful physically but once we get to know that person we discover such beautiful feel-ings that that person becomes beautiful to us. So that beauty is a feeling inside us and when we discover it, it makes us see beauty where some others don't. It's a feeling.

What do people say in your community?

I think that people are guided by what society deems beautiful, the media says that being young and having a motor bike is beauty. Mainly being young, that is what sells. Then they provide us with some beauty parameters that are false, for me, completely false. I do not believe what society says about beauty, it is only for sales purposes.

What do you see as your connection, if any, to the people living in Morocco?

There is a common history, Arabs spent 800 years in Spain, eight centuries, and they left us many things from their customs, beliefs, architecture, culture . . . so I think that there are strong connections between us. Culturally speaking, we owe the Arabs a lot; that is the Moroccans and the Arab culture. I think that we can list more things that link us than things that separate us.

What are your biggest differences?

I think that the main difference is that they still are an underdeveloped country while we are an industrialized country. I think that there is an economic difference that makes us different. They have some needs that we have already covered. Apart from this, I think that there are no big differences. They may have the same needs that (we have) but in different levels. They live in more precarious conditions than we do in relation to all levels, the economic level, the political level . . . they still have to achieve a democratic government in order to live in freedom, something that we finally have after many years and efforts.

Would you like to go to Morocco?

I'd love to.

Why?

Because I think that it is a fascinating country, a country of contrasts, of light. I think that there we could find the roots of the Andalucian people because we share many things with them. It is one of my greatest challenges, to go there and . . . I hope it will be soon, to go there for a reencounter, because I think that it is a reencounter, a reencounter with a culture we are part of, in some way. I think that it is something that we all should do, all the Andalucians, at least once in our lifetime we should go to Morocco and meet our ancestors.

What do you think about immigration and the problems that there are now in relation to this?

Yes, they are doing the same as we Spanish did in the 1960s. Due to economic reasons and shortages, we immigrated to Europe, mainly to France, Germany, Belgium, Holland . . . looking for a better way of living. This is the same as what is taking place now. The Moroccans and many people from other parts in Africa are trying to enter into Europe to improve their lives. In their homelands, their lives are in jeopardy, they are dying of hunger, so they have the right to look for a place where they can fulfill their aims in life. What happens is that we normally forget that some time ago we were also immigrants, we have forgotten about that. So we should give refuge to them, look for the best way of helping these people to find a place to live, even to survive, because they really are starving to death. We should express more solidarity. Spain cannot become Europe's police and just close the doors.

Carmen Barragán Diaz, 42, lawyer, Seville[1]

Jamal Rhmiro is an educator in Fes and like Carmen, an extraordinarily thoughtful person. Jamal's anger toward the Spanish regarding their treatment of Moroccans living in Spain is eloquent and illuminating. Like Carmen, Jamal sees the commonalities between Spain and Morocco that many Moroccans would acknowledge if pressed. Jamal is merely openly stating what many others are thinking. This is the level of honesty that will be required if Spain and Morocco are to reconcile in a fundamental way. Anger must be given an open space to be aired. Jamal blends his anger with a faith and a willingness to move ahead. Only after this has been done can the work of bridge building begin. Finally, Jamal's acknowledgment that often the governments do not speak for the people or take their situations into account lends credence to the precepts of Sustainable Diplomacy. His call for a new level of communication points to the possibility of a new type of relationship.

Jamal Rhmiro

How do you feel connected to the land where you live?

This feeling [of connection] in the city is dying little by little, because a connection to the land does not exist. The relationship with the land that feeds us does not exist [here]. In the city we see roads and walls so the connection with the land is lost. This club [the club of teachers] has trees. Here you feel that land exists. In other places, you forget that there is land.

Can you say where this comes from?

Because we have neglected the rural areas where this relationship exists. Here [in the city] you can go for a year without touching land. You can only get this all back when you go to the rural areas, where you can see that the land and nature exist.

Do you own any land?

Yes, I do. It was my grandparents' and it is in a rural area.

Do you think that your job is related to the land?

Yes, because we are all related to the land because of gravity. We cannot work if our feet are not on land. It is even more the case as I am a P. E. teacher. The more you work on the land, the more it will give you. If we have the opportunity to help the people who will become farmers, our job could help them to work the land more, meaning that if you don't have good health, you won't be able to take care of the land.

How is your life affected if the rain does not come?

If there is no rain, I feel deceived—a feeling I cannot explain. Whenever there is rain I feel very happy.

What is your role in helping the coming generation?

I'm not going to tell you that I'm an altruist . . . . I think that if you prepare the coming generation, you prepare your own future. We are getting older, [so we] can find [our former students] in medicine, in architecture, in all the helping professions. So you prepare a child so he can welcome you at some time in the future. So you have to prepare him to inherit this land and this world. The inheritance isn't what you give him, but rather what he is going to give to this land, so you should prepare him to do that.

Do you think that their future is connected to the health of the land?

Yes, it is important, but will it be direct or indirect? Recently it has become indirect, because civilization has swept [away] this relationship. But this connection will always [have the potential for being] direct. I think that the coming generation will feel it more because the pollution will be more and . . . . they will stand against it as in the western countries, and all the entertainment will become oriented toward the countryside—toward land.

What does your religion and tradition teach you about the origin of the land and how it was created?

As Muslims, [we believe that the] land was created by God's will, and not the Big Bang, as they say. An ayat in the Qur'an says: "He created the skies and the land in six days and then He sat firmly on the throne." The land was created for a specific reason.

Can you say where you learned this from?

First from my family, secondly from school, thirdly from the street and then from personal experience.

Will you teach these things to your children?

Sure. There are some specific principles you want to pass on to the children. This is where education lies.

What does your religion and your traditions teach you about the relationship between land and other creatures?

The Islamic religion obliges you to respect anything you have a connection with, be it human beings or other living creatures. An example of this is that Islam obliges you to, before touching anything, to say "In the name of Allah" in order that you enter into a peaceful place between you and what surrounds you.

Does your religion or tradition oblige you to save other creatures?

Yes. As an example, the Prophet said that when you want to perform the ritual ablution, you must use a minimum of water. This is one way to take care of water and save it. I think that the Prophet said that if you want to perform the ritual ablutions, and you took water from a river or a sea, after finishing you should return the remaining water to its [original] place. I think that there is no other example greater than that.

In your religious tradition, is there a way to speak about the beauty and the holiness of the land?

The land or the universe as a whole? I think that the greatness of God is shown only in very small things. You could go to paradise if you just contemplate [them] and cry.

Is there a way to understand God's greatness in God's creatures?

As I said, you can take the simplest beings and they will show you that greatness; starting from life as a whole—how the universe is moving with unlimited exactitude, to the most minute thing, which is moving with the same precision.

What do people say about the greatness of God?

People in the past have said there is a feeling connected to the greatness of the Creator which shows in our lives. By reflex, we say: "Greatness to God, Glory to God," something that comes out in our daily lives without our ever thinking about it.

How do people in your community speak about the blessings that come from the land?

Anything that comes from the land is a blessing that we thank God for, thanks and praise to God.

How do you look to the Spanish, our neighbors?

They are a people who seem to be developed, but only at first glance. They have not truly developed in a deep sense. This comes from an inferiority complex. They were like us, but they became better, but we remain as they were. If you look at them they are more like us than they are the French or the English. They are not well accepted in the EU. Their behavior is bad. I may be influenced by the media, but when I see a young Spaniard beating an old Moroccan man, I start to think about many things, and I begin to look at the Spanish as inferior.

Do you think Spaniards are different than Moroccans?

No, I don't think so. They simply had the opportunity to become developed. But as I told you, their development is only superficial. They are not actually different.

If you got a chance to travel to Spain, would you go?

Yes, I'd like to spend my holidays there. My vision [of Spain] has come through the media, but I would like to [have] contact [with] them and see their reality.

Can you say what you think is the most important thing to do to improve relations between Morocco and Spain?

People, be they Spanish or Moroccan, shouldn't look at things as simply as they have. For example, when a problem occurs between Morocco and Spain, its better not to be limited by what others say, or what the Moroccan or Spanish media say. We should think about things. There are some decisions that came randomly. The governments do not have the same worries as their people. There are decisions made on a government level, which do not take the people into account. We should find a means of communicating, which brings more understanding.

Jamal Rhmiro, 35, teacher, Fes[2]

Reading the Sacred Texts: Abraham and Ibrahim and the Strangers

Both Carmen and Jamal are familiar with the experience of being viewed as "other." As a feminist in southern Spain, Carmen is not a typical

Andalucian woman. She has, however, found through her own "outsider" status compassion for others who are seen as "different." She has made a critical connection—one that must be made by others if Spanish society is to be transformed into a more welcoming place for immigrants. Carmen can welcome North Africans into her country because she can see her own country's story in theirs. Jamal shares the same ability, but for different reasons. Having had the experience of travel in Europe, Jamal knows what it is like to find himself as the outsider. His command of multiple languages enables him to negotiate a place for himself in his travels. This is not possible for the often far-less educated groups who cross the Strait illegally in search of work in Europe. Through television, Jamal has seen what can befall a Moroccan in Spain. His reference to a Moroccan man being beaten is a reference to an anti-immigrant riot that took place in the southern Spanish village of El Ejido in February 2000. While the events in El Ejido were exceptional, the images that came onto Moroccan televisions of this event had a wide and lasting impact across the country. Understanding and accepting the "other" is therefore a critical theme in any discussion of Moroccan–Spanish relations. The theme of the stranger being welcomed is central to the story of Abraham (or "Ibrahim" as he is known in Arabic), a story shared by the Bible and the Qur'an. And so we turn to examine these texts for lessons in hospitality and inclusion.

From the Bible

The Lord appeared to Abraham by the oaks of Mamre, as he sat at the entrance of his tent in the heat of the day. He looked up and saw three men standing near him. When he saw them, he ran from the tent entrance to meet them, and bowed down to the ground. He said, "My lord, if I find favor with you, do not pass by your servant. Let a little water be brought, and wash your feet and rest yourselves under the tree. Let me bring a little bread, that you may refresh yourselves, and after that you may pass on— since you have come to your servant." So they said, "Do as you have said." And Abraham hastened to the tent to Sarah, and said, "Make ready quickly three measures of choice flour, knead it, and make cakes." Abraham ran to the herd, and took a calf, tender and good, and gave it to the servant, who hastened to prepare it. Then he took curds and milk and the calf that he had prepared, and set it before them; and he stood by them under the tree while they ate.

They said to him, "Where is your wife Sarah?" And he said, "There, in the tent." Then one said, "I will surely return to you in due season, and your wife Sarah will have a son." And Sarah was listening at the tent entrance behind him. Now Abraham and Sarah were old, advanced in age; it had ceased to be with Sarah after the manner of women. So Sarah laughed to herself, saying, "After I have grown old, and my husband is old, shall I have pleasure?" The Lord said to Abraham, "Why did Sarah laugh, and say, 'Shall I indeed bear a child now that I am old?' Is anything too wonderful for the Lord? At the time set I will return to you in my due season, and Sarah shall have a son." But

Sarah denied, saying, "I did not laugh"; for she was afraid. He said, "Oh yes, you did laugh."

—Genesis 18:1–15

Let mutual love continue. Do not neglect to show hospitality to strangers, for by doing that some have entertained angels without knowing it. Remember those who are in prison, as though you were in prison with them; those who are being tortured, as though you yourselves were being tortured.

—Hebrews 13:1–3

From the Qur'an

Our angels came to Abraham with good news, and said: "Peace on you." "Peace on you too," said Abraham, and hastened to bring a roasted calf. When they did not stretch their hands toward it he became suspicious and afraid of them. They said: "Do not be afraid. We have been sent to the people of Lot." His wife who stood near laughed as We gave her the good news of Isaac, and after Isaac of Jacob. She said: "Woe betide me! Will I give birth when I am old and this my husband be aged? This is indeed surprising!" "Why are you surprised at the command of God? God's mercy and blessings be upon you, O members of this household," they said. "Verily God is worthy of praise and glory."

—Hud 11:69–73

Inform them about the matter of Abraham's guests. When they came to him and said: "Peace," he answered: "truly we are afraid of you." "Have no fear," they said. "We bring you news of a son full of wisdom." "You bring me the good news now," he said, "when old age has come upon me. What good news are you giving me then?" "We have given you the happy tidings of truth," they replied. "So do not be one of those who despair." "Who would despair of the mercy of the Lord," he answered, "but those who go astray."

—Al-Hijr 15:51–56

So We gave him the good news of Isaac, apostle, who is among the righteous. And we blessed him and Isaac. Among their descendants are some who do good and some who wrong themselves.

—As-Saffat 37:112–13

Common Ethical Principles: The Lesson of Abraham/Ibrahim and the Strangers in the Biblical and Qur'anic Accounts

1. Offering hospitality to strangers is not optional: it is the responsibility of all Christians and Muslims

Christianity and Islam share many things in common, including a Near Eastern origin. In the Near East, the obligation of hospitality arguably preceded the birth of Judaism. As in many regions of the world where people have lived in scattered and often inhospitable terrains, the notion of giving hospitality was a central cultural obligation. Without this tradition, many a traveler would have perished.

It is only logical, then, that hospitality plays a prominent role in both Christianity and Islam. Abraham's encounter with "the three strangers" is emblematic of a long-standing social expectation that became a religious obligation. For the Christian, Abraham sets a standard that is echoed throughout the Hebrew Bible and the Christian Testament. Abraham's actions are repeated by many other Jewish prophets up through and beyond the teachings of Jesus. In many instances, we find that accounts of hospitality are used to illustrate deeper truths; of expectations thwarted, Divine intervention, and the folly of prejudice revealed (the story of the Good Samaritan and the story of the road to Emmaus are just two examples). It is for this reason that giving one's own possessions to a stranger in need remains one of the basic elements of Christian faith and practice.

Islam has the same high regard for hospitality toward strangers. In keeping with basic Bedouin ethical standards, the Muslim tradition of offering hospitality and protection to strangers emerges from very similar circumstances to those of Judaism and Christianity. Many Muslim cultures have defined themselves in this free offering of a lifeline to an unknown person. The Qur'an and Hadith are replete with examples of such kindness. The theologian Ibn al-'Arabi echoes this sentiment when he writes:

> It was because Abraham attained to this rank by which he was called the Intimate [of God] that hospitality became a [sacred] act. Ibn Masarrah put him with Michael [the Archangel] as a source of provision, provisions being the food of those provided. Food penetrates to the essence of the one fed, permeating every part.[3]

Ibn al-'Arabi makes the extraordinary claim that when one offers someone else food, that one is offering a part of Creation itself—something that cannot be separated from God. Thus, the giving of food is literally sharing the Divine with another.[4] The Christian tradition of sharing the bread and the cup reflects the same sentiment in both of these simple, vital acts, one of the most important essences of each faith is revealed for those who would receive it.

2. *It is in encountering the one we call "other" that we recognize the truth about ourselves, and our common Divine origin*

For both Muslims and Christians, encountering the stranger is a defining act. The beliefs and subsequent ethical standards of each community are revealed. For Islam, the beliefs and practices of the *umma*, or global Muslim community, are reflected in the encounter with a stranger. In the opportunity to show mercy, kindness, and generosity the Muslim community finds itself reflecting the teachings that the prophet Muhammad, with the help of Allah, was able to live out. Likewise for the Christian, acts of charity toward the stranger are opportunities to exemplify the teachings of Jesus and his community of followers.

The story of Abraham and "the strangers" is one that invites both Muslims and Christians to question their suppositions about those unknown to them. While both traditions remind their followers that all humans were created by God and reflect the Divine in Creation, such claims are more easily heard than acted upon. It is for this reason that the Qur'anic and Biblical accounts of Abraham and his response to the arrival of strangers at his tent is more than a mere story. It is an admonition to resist the temptation to separate along lines of ethnicity, class, race, and religion and to share only with those of one's own identity. The truth is this: one cannot know at the first or even second glance whom one is actually dealing with, let alone their origin or their purpose. The ancient story challenges Muslims and Christians to see how far they are willing to go in acting on their claims of belief. Their level of faith will determine if they can see God in the eyes of someone they have never met.

3. Offering sustenance to strangers can sometimes push us to the edge of our comfort level. What we have not done before is often the most difficult thing to do of all

In the Qur'anic and Biblical accounts, Abraham and his wife encounter "the strangers" with varying and sometimes uncomfortable responses. In the Biblical account, we are told that Sarah "was afraid" (Genesis 18:15), when she heard the prophecies of the unexpected visitors. The Qur'an tells us that Abraham was leery of what his unexpected visitors had to tell him. The eigth-century text *Stories of the Prophets* affirms this when it recounts that "Abraham was . . . not so sure about their [the angels'] glad tidings at this old stage of his life."[5]

For Abraham and Sarah, the tables are turned. The initial "givers" of a meal become the recipients of a miraculous prophecy that they will finally have a child, something that they had thought impossible. An encounter that was already unusual became stunningly unfamiliar and even frightening. Entertaining any stranger introduces an unknown factor, and encountering the Divine in the unknown person is not necessarily a settling experience. This is the risk in offering help. The nature of the risk, however, remains as unknown as the person who brings it, until some unexpected surprise presents itself. How we respond in this moment may define us and our destinies. When Abraham and Sarah served a meal despite their reservations, they became the model for Christian and Muslim conduct.

4. The world that humans inhabit and the realm of the Divine are constantly intersecting, producing challenges, beauty, and fear while defying all expectations. The stranger we encounter could in fact be God or an emissary of God

Muslim and Christian believers share the understanding that human life takes place within what Augustine of Hippo described as the *saeculum*. The saeculum is the place where humans live out their earthly lives, a point of intersection between the world of the Divine and the limited world of the human.[6] Living within this confluence demands important decision-making on a daily basis. One can therefore choose the option to participate in the life of Creation as God intended, or thwart such a design and risk the loss of one's salvation. As times change, the distinctions between these two paths, which appeared so clear to many earlier followers of Christianity and Islam, are now more difficult to discern.

As we saw in the Islamic symbol of the inverted Tree of Being meeting its infernal counterpart, the followers of God live in a world of daily potential confusion and temptation. The Christians also live in such circumstances, many of whom are taught that the world they live in is suffused with both good and evil. Making the choice to "do good" as Abraham and Sarah did in feeding their mysterious visitors is thus portrayed in both traditions as opting to be on the side of God. Abraham and Sarah made this choice despite any doubts they may have had as to the origin or purpose of their visitors. Their choice was one of faith followed by action. Their risk was in turn rewarded by a marvelous and all but unbelievable prophecy: of the birth of their child. Abraham and Sarah are invited to suspend their earthly disbelief. Clinging to their accustomed sense of reality would have robbed them of an extraordinary gift. "Why are you surprised at the command of God?" said one of the Qur'anic visitors (Hud 11:73). "Is there anything too wonderful for the Lord?" echoed one of the Biblical visitors (Genesis 18:14). We do not always know the true nature of the visitors we entertain; yet we will never know anything about them unless we extend a sincere invitation to them to come to our table.

5. *Extending help to strangers is a concrete expression of practical faith and solidarity, for today's host does not know when he or she will be forced to rely on the compassion of strangers tomorrow. Conversely, those who are recipients of aid must also be people of faith and be open to receiving what they are being offered*

Expressions of faith and practicality need not be mutually exclusive. The Bedouin ethic of hospitality emerged out of necessity, acknowledging that all people are vulnerable travelers at some point in their lives. Thus one must become capable of being both a generous host and a gracious guest, lest the system of survival fall apart. This was arguably the initial drive behind Abraham and Sarah's act of hospitality. They too had been travelers and would likely be so again. Yet the social arrangement this exchange describes is hardly limited to a *quid pro quo* sort of transaction. This gift freely given by an elderly couple was a sign of their faith and goodwill, reflecting the extraordinary confidence that God had placed in them.

The ritual of sharing a meal with a stranger becomes part of a larger cycle. It potentially changes the circumstances and perceptions of all participants. The giver has made an investment of time and wealth while risking potential harm. The recipient, knowing nothing of his or her host, enters into a space that is initially unknown and potentially uncomfortable or even dangerous. Both sides place faith in the goodness of the other.

Both sides bring their own respective faiths to the table. Eating a meal together is among the greatest of levelers, because it acknowledges our common needs and dependencies. To share a meal is to share the gifts of Creation, consciously or unwittingly. To decline to participate in such an experience is to potentially miss out on the possibility of a powerful physical and spiritual transformation. Abraham and Sarah and the guests said "yes" to the possibility of transformation, and no one was disappointed.

Points of Disagreement: The Lessons of Abraham/Ibrahim in the Biblical and Qur'anic Accounts

1. For the Christians, Abraham was the first Jew, and for the Muslims he was the first Muslim

Muslims and Christians lay their own particular claims on Abraham. For the Christians, he is the first member of the people that would become Israel: the first Jew. Abraham's faith in God extended even to being willing to sacrifice his son at God's request.[7] For the Muslim, Abraham is the first *hanif*, or monotheist. For this reason, he is most often considered to be the first Muslim, for in Islam's eyes Abraham predates the arrival of Judaism. Therefore, Muslims, like Christians, trace the beginning of their sacred family tree to Abraham, and see him as the founding member of their tradition.

The questions of identity and tribal affiliation that serve to separate Christianity and Islam are present in the stories of Abraham. In confronting these differences, Christians are given a unique opportunity to experience the same confusion Jews often grapple with in encountering conflicting Christian claims of origin and meaning. The names of figures are similar but their roles have often changed. For the Muslim, however, the revelation of the Qur'an is seen as the ultimate clarifier of terms, identifications, and affiliations. It is the last revelation of God. For the Muslim, Christians and Jews are "Ahl Al-Kitab" or "People of the Book," and Islamic identity is most prominently linked to its predecessors via the Qur'an and its mention of the intrinsic relationship among the three "Abrahamic" traditions. Textually, one could argue that Islam is more inclusive than Judaism or Christianity, based on Qur'anic passages.[8] Christianity does not generally extend the same courtesy to Islam, which it traditionally views as a heretical tradition, given that Christians believe Jesus of Nazareth was the embodied manifestation of God on Earth. For the Muslims, while Jesus was a great prophet, he was still an earthly man. While these competing and contradictory claims are not easily reconcilable, they point to the place upon which the builders of bridges between the two traditions must set to work.

2. Many Christians and Muslims are equally wedded to the notion that their faith is the exclusive source of salvation in the world

One of the greatest stumbling blocks for Christian and Muslim cooperation on the ground is the belief in an exclusionary principle: a great number of believers claiming for themselves an exclusive hold on God's salvation. This is a fact that is often downplayed among those who seek to fully reconcile Christians and Muslims. Yet long-term viable cooperation will never be realized until this disparity is directly addressed.

While such differences seem obvious, it is remarkable how seldom they are addressed head on, save for one side or the other attempting to convert the followers of the "other" tradition. The time has come for all Christians and Muslims to ask themselves how closely they will choose to cleave to the often popular exclusivist readings of their respective faiths. There are regretfully few Western traditions to look to for examples of an inclusive soteriology that recognizes salvation in the terms of another religion. It appears that it is often only in the realm of mysticism that the Abrahamic traditions can transcend culturally bound exclusivist claims of their respective faiths. Sufi theology is for this reason one of the most fertile grounds for breaking the molds of both Christianity and Islam. Combined with the work of Christian mystics, both modern and ancient, there is an opportunity to rethink the claims of popular Christianity and Islam. The "theology from below" that Dietrich Bonhoeffer spoke of, that of the theology of the oppressed, is clearly another critical source for both Christians and Muslims who strive to build a new type of dialogue on a deeper and more meaningful level.

Regretfully, it is often the most popular interpretations of Christianity and Islam that are the least inclusive. This is exactly why those who believe in countering the exclusive claims of the popular readings of their faiths must learn new ways to speak out across multiple boundaries in a manner that is compelling and intelligible. One of the clearest sources for a new inclusive theology is found in the many theologies of liberation emerging in both Christianity and Islam. Such theologies place special value on the life and experiences of the most marginalized peoples and thus speak to many people across current lines of difference. Liberation theology challenges its adherents to fight for the realization of new visions, which other traditions would term idealistic, if not impossible. Liberation theology sees the persistence of poverty as a sign of sin, racism as an expression of the betrayal of God's design, and sexism as an insult to the work of those who strive to follow God. At their best, liberative theological approaches turn accepted realities on their head, while they creatively confront the obstacles that prevent many people from recognizing that all humans are equal in the eyes of God. This theological work is being carried on today in the work of Farid Esack, James Cone, Gustavo Gutierrez, and many others. Mainstream theology alone will not forward the project of Sustainable Diplomacy: it will more often than not ensure its failure.

Pushing the Limits of the Ecological Footprint: The Commodification of Illegal Human Labor in Spain

Thus far, the Ecological Footprint has provided no means to calculate a nation's use of human labor, legal or illegal. While it calculates the hectares of different types of land, the Ecological Footprint stops short of measuring the number of human laborers necessary to cultivate, process, package, distribute, and market goods and services for the general market. Those who are working to expand the scope of the footprint, such as Chad Montefreda of Redefining Progress, explain that human labor is omitted because the focus of the Ecological Footprint is the use and consumption of nonhuman materials.[9]

The Ecological Footprint remains highly useful in analyzing and comparing the consumption patterns of individuals, cities, regions, and nation-states; but it fails to tell the whole story. In particular, it omits the costs connected with the exploitation of human beings who have entered a country illegally to seek work when none is to be found in their own homeland. There are many dimensions that could be explored in this area, including: (1) economic dependency on cheap illegal labor; (2) the diminishment of a nation's Ecological Footprint through the use of an illegal population that is unable to consume at the same high rate as their legal neighbors; (3) the percentage of a host nation's bio-capacity, which is protected or consumed by the use of illegal labor; (4) the level of energy that is put out by illegal labor and in turn harnessed by ostensibly legitimate businesses; (5) the fact that illegal labor conceals the true costs of the commodities produced, potentially skewing the Ecological Footprint of a country toward a lower number of hectares per person than is actually the case; and (6) the loss of productive capacities by the countries from which the people emigrate.

One of the greatest crises currently facing Spain and all of its sister EU countries is the issue of illegal immigration. The crisis plays out on a number of fronts, including issues of mobility, racism, employment, housing, religious tolerance, and the very economic future of western Europe. With a legal labor pool of approximately 17,000,000 people, Spain also hosts an estimated 200,000–800,000 illegal laborers, a great many of whom are Moroccan.[10] While estimates clearly vary, they all may be lower than the true figure.

Spain's growing dependency on illegal labor can be directly traced to its low birthrate of 1.1 children per woman, the lowest in the world.[11] As Spain's birthrate has shrunk, so too has the portion of its native population who are willing to work in the agricultural sector, as many Spaniards have migrated to the cities over the last 20 years. These combined phenomena have left Spanish farm owners with few alternatives other than foreign labor to bring in their crops. In the Andalucian region of Almeria alone, an estimated US$1.8 billion of fruits and vegetables are grown annually. Until recently, most of the people participating in the harvest were Moroccan. Most unfortunately, the large size of the Moroccan population of Almeria has led to increasing tensions between the Muslim and Christian populations.

These tensions exploded in the Almeria town of El Ejido on February 5, 2000. The spark that ignited the ensuing riots was the murder of a local young Spanish woman by a Moroccan who was later described as "mentally disturbed." The resulting violence showed that the views of some Spanish residents of El Ejido are connected to long-held resentments and prejudices:

> Residents, who had evidently planned their revenge, blocked off roads with barricades of burning tires to delay police officers and television crews.
>
> "They came with sticks and bars," said Kamal Rajmuni, a Moroccan community organizer in El Ejido. "They always treat us like dirt, but this time they behaved like Nazis." Many people had to run for their lives. Witnesses said bands of Spanish men, many of them young, screamed, "Out with the Moors!" as they threw stones at the foreigners. Several Moroccan teahouses and two makeshift mosques were destroyed, and firebombs were thrown at immigrant shanties.
>
> The violence shocked many Spaniards . . . People in this country had long been used to reports of terrorist acts by Basque separatists in the north. But they had never before seen television images of immigrant hovels burning or small stores destroyed in Andalucia.[12]

While the violence in El Ejido is by no means the typical Spanish response to the presence of foreign immigrants, it demonstrates what is possible when people perceive that they have gone beyond the limit of their own capacity for tolerance.[13]

Spain and Illegal Immigration in the Greater Context of the EU

There is growing tension in Spain because of the role its EU partners have asked to play in response to the increasing presence of foreign immigrants, particularly those who are North African. Since the beginning of the expulsion of the Muslims in 1492, Spain, like many of its neighbors, has in many ways become an ethnic and religious monoculture. Despite many Spaniards' partial African heritage, most Spanish citizens identify themselves as purely European in origin. In turn, if they identify themselves as religious, it will almost always be as Christian. With the exception of such large multicultural cities such as Lisbon, Barcelona, Rome, Paris, Berlin, and London, the same could be said about the bulk of western Europe until recently. This situation has left a broad and deep cross-section of Europe's citizens having no substantive experiences in interacting with people of other races or religions. A town such as El Ejido, which prior to the riots had a population estimated to include over 100,000 North Africans, was an anomaly—and to many Spaniards a threatening one.

It is currently estimated that there are at least between three and five million illegal immigrants living in the EU.[14] The significant growth rate among illegals has led to some troubling developments. For example, when the so-called minority grows to a critical point, a different mind-set among the "native" majority population often emerges. In the face of a potential power shift, some European inhabitants feel threatened. The increasing popularity of "anti-immigrant" political parties across Europe attests to this fact. Whether the parties are based in Spain (Democracia Internacional), France (The Popular Front), Britain (The National Front), Italy (Northern Alliance), Denmark (People's Party), the Netherlands (Leefbar Nederland), or Austria (Freedom Party), their appeals are all similar: there are too many immigrants in Europe, and their numbers must be cut at any cost.

One of the results of the increasing strength of the right within the EU's membership has been an elevated level of pressure on Spain to stem the tide of illegal immigrants entering Europe from Morocco via the Strait of Gibraltar. Spain has thus become one of Europe's "front line" nations in the war to keep illegals out. The Spanish response to this call has been multifold. Spain has installed a US$150 million electronic warning system composed of a "350 mile electronic wall to prevent illegal immigrants from entering Spain."[15] This "Integrated Electronic Surveillance System combines radar, infrared sensors and night sights to provide a blanket coverage of the Spanish coastline" from Huelva in the southeast to Almeria in the southwest.[16] Spain has also significantly increased its marine patrols in the Strait, particularly at night. Finally, an EU jointly funded project has erected five miles of parallel running fence around the Spanish enclave of Ceuta at a cost of over US$25 million. These fences are monitored by 33 closed-circuit cameras and sensors in an effort to detect any human presence near the parameter.[17] These combined efforts have resulted in the apprehension of as many as 324 migrants in one single day.[18] The number of those who evade Spanish detection is difficult to measure, as is the number who perish in the overcrowded boats, which some passengers are charged as much as $2,500 to ride in.[19] Many boats, packed far beyond their capacity, sink in the Strait's treacherous waters long before they reach Spanish soil. The hundreds of bodies that wash up along the shores near the Spanish town of Tarifa attest to this fact. All the barriers in the Strait have forced migrants to focus on entering Spain via its southernmost point, the Canary Islands, off the coast of Moroccan-controlled Western Sahara.[20] Whether by the Canaries or through the Strait, one factor is constant—the participation by Spanish syndicates who profit off the smuggling of illegals.

These illegal immigrants are not only citizens of Morocco. People come from as far away as southern sub-Saharan Africa to take their chances on Spain's well-guarded borders. Many who are caught tear up their identity papers in an effort to stymie Spanish efforts to repatriate them to their country of origin. Because Morocco is the favored passageway of many Africans, the EU provides economic incentives to Morocco to stem the flow. One project initiated in 2000 by Spain, called the PAIDAR, financed incentives for Moroccans to remain in some of the poorer northern regions of Morocco.[21] These efforts also included Spain making an effort to help Morocco to eradicate its significant cannabis crop, which may be the greatest source of hashish smuggled into the EU, and to replace these crops with legally marketable fruits and vegetables. Given the market price of hashish versus tomatoes, the potential success of such a project is questionable.[22]

Ultimately, Europe is left with the same dichotomy as the United States: in the absence of foreign legal and illegal labor, many regional economies would be dealt huge blows and perhaps even collapse. Increasingly, manual labor jobs previously held by European nationals no longer hold any attraction for the younger generation. This is easily demonstrated in Spain, which, despite having one of the EU's highest unemployment rates, is unable to attract Spaniards to work the harvests.[23] This is not surprising, given that a legal worker might earn just US$300 a month.[24] An illegal worker earns far less. For this reason, many Moroccans, Ecuadorians, sub-Saharan Africans, and others working the harvests live in shanties without electricity or running water.[25]

Another phenomenon exacerbates anti-immigrant tensions. As many Muslims reach Europe, they often draw closer to their culture and their faith, living as they are in frequently unfamiliar and incomprehensibly difficult circumstances. Simple practices emerge from this phenomenon that would not normally draw notice. A woman wearing a head scarf in a small village stands out. People who gather to practice a different religious tradition stand out. Those who don't speak the native language are marginalized. And when new arrivals do learn to speak the language, it is with an accent. Speaking Arabic in public is often frowned upon. None of these phenomena are a legitimate excuse for discrimination, but many Spaniards (and others) will voice their resentment at those who in their eyes "do not try to blend in."[26]

On the other side of the coin there are active antiracist organizations throughout Europe. In Spain the Movement Against Intolerance has been the source of numerous legal briefs on behalf of illegal workers, specifically Moroccans.[27] In addition, a recent immigrant-rights protest movement based originally in French churches has spread to Spain. In May 2001, 329 Moroccan and Algerians seeking legal immigrant status went on hunger strike in a church in the southern Spanish town of Huelva and garnered a significant amount of support from local Spaniards.[28] Another example of Europeans advocating on behalf of immigrants are the Italian Missionaries of St. Charles, also known as the Scalabrinians, who work to assist migrants. Today this organization works "in 25 countries operating migrant assistance centers."[29] Finally, one cannot discount another interesting phenomenon: the small but growing conversion to Islam of European and especially Spanish citizens. These groups, most often Sufi, are still small. Yet these religious communities signify a new type of openness that could not have been predicted 20 years ago.[30]

The Response to Immigration: Spaniards and Moroccans Speak

What do you think would be the most important thing to do to improve Morocco's relationship with Spain? Imagine that you are the king of Morocco or José Maria Aznar and that you can change whatever you want in order to make these connections better.

I think that . . . it is right if the Spanish people go there to work and the Moroccan people come here to work too. I think that there shouldn't be frontiers and that the Spanish people who want to go there to work should be able to do it and that the Moroccans who want to come here to work should be allowed to do it. Because, here, we need people to work and if they have to be from other countries . . . When I was a child, when I was about 14 years old . . . after being a shepherd and all that . . . one of my uncles took me to the country to collect olives . . . I've been working since then.

Juan José Vardera Luis, 74, olive farmer, Quatravitas[31]

What do you think of the Spaniards?

Thanks be to God, God has created races equal to one another, the only difference lies in religion, language, and geography. There is a line that separates, so there should be a strong relationship between us and our neighbors, the Spaniards. They have a humanistic soul. Arabs lived in Spain, Moroccans lived there, and built monuments there, and they are proud that there are Arab monuments. In fact they love Moroccans, and Moroccans have a humanistic spirit, and they love Spain. As long as there is cultural exchange, because we belong together to the Mediterranean basin, there is not a long distance between us. Also, our civilizations are close. The lifestyle between Moroccans living in the north of Morocco is the same as that of Spaniards living in the south of Spain. Here starts a knowledge point that must be complete and global, but there are some circumstances that create problems between us. But if Moroccans and Spaniards have a strong spirit, and are free from hatred, relations will be good. There is also another issue, which is immigration . . . . Personally, I think that there should be coordination and cooperation in many fields, including economics and exchanges of manpower. They need farmers, and we need highly qualified cadres, and the consolidation of the sports sector. I suggest they help us with technicians so we will have champions who will be distinguished in many fields. Also, infrastructure and facilities should be built, as well as investment between us. [The Spanish should] build factories here, and avoid immigration, and save the lives of those immigrants who go to Europe. The Moroccans will find places to work here, and think that immigration is wrong, and the same for [the Spanish]. Illegal immigration then will be limited, and Morocco will then not be accused of being the starting point of this type of immigration.

"Sidi X," 45, potter, Fez[32]

What do you think you have in common with the Moroccans?

I personally don't like the Moroccans, because they are false. They are liars and steal from each other. I have never been to Morocco, but when I [worked] in another hostel in Calle Santa Ana, there were many of them and I couldn't cope with them. They were unbearable. And one day, they were complaining about things that were missing and that we had stolen them, and behind a mirror there were 6 or 7 passports that were stolen from other people. I don't know, I don't like them. They come here to Spain and steal a lot, I have seen them doing that to many people.

I saw a couple from the USA coming out of the metro, and one of them got in front of the couple and the other one behind them. Then one of them [acted] as if he was falling and the couple couldn't get through. Meanwhile, the other stole their wallet. It made me really mad. So, ... I have never seen a Spaniard steal, although there are all kinds of people everywhere. But here in Spain they [Moroccans] steal a lot. Almost every night we have a client here that has been mugged and it's always them. I don't know if they are Algerians or Moroccans, but they are from that part of the world. They are dark and always the same. At the same time I don't think that they are all bad. There must also be good people, but I have this image. I don't know exactly where they are from, but as you have been there, you might have another concept. And I respect that from you as well, don't I? Because it also appeals to me that you have another idea about them and that it's a better one than mine. You have been there and got to know them, but for your job, a different system, you didn't have to fight with them in a hostel and they didn't make you angry. With other people you can have a bad day, but with them [it's a bad day] continuously.

Maria Concepcion Sara, 59, hostel desk attendant, Barcelona[33]

What can be done to improve the relationship between Morocco and Spain?

In my opinion Spain looks upon Africa and the Arab and Islamic world as undeveloped regions. They see themselves as the most intelligent people in the world. They see themselves as having industrial [developed] regions and so on, despite the fact that when watching a football match they [often] perform vandalism. They are not the most industrious people in Europe. The industrious people are but a few and they are in the laboratories. [Many Spaniards] see themselves as having a civilized nature, but what I hope is that they get rid of their xenophobia. This xenophobia is a sort of racism against people coming from abroad through Morocco. These people have suffered from oppression. For example, Moroccan products going to Europe are attacked [when they arrive] in Spain. These things become known to everyone and they leave an impression. So they have to get rid of such behavior and leave Morocco in peace. They are attempting to create problems in Morocco. They colonized Morocco and went away, but it is clear that they have created problems among Moroccan citizens that persist after they finally left. And if our citizens keep on struggling with these problems they will never develop. And it has become clear to everybody that other countries want us to have problems among ourselves so as to [prevent] us from developing. I hope they get rid of this behavior. They say that they are civilized, but what they are doing has no relationship with civilization at all. It has nothing to do with development or civilization.

Abidine, 26, science student and jewelry maker, Guelmim[34]

What do you think would be the most important thing to do to improve the relationship between Morocco and Spain?

I'm not racist, I don't care if a person is Black, Muslim, Spanish, or French as long as he doesn't try to cheat me. I used to get on well with them and vice versa. They may cheat me because I let them, because it is to my interest . . . but I'm not stupid . . .

I know good and hard working Muslims, who work even more than the people from Spain. Here, I have (a Spanish) employee who is totally useless while there, the Muslims are waiting to work from six in the morning until sunset and they are paid almost nothing. That has happened to me in Ceuta.

Antonio Sala Rodriguez, 35, custodian, Seville[35]

What do you think should be done to improve relations between Morocco and Spain?

We should treat one another as neighbors, each should perform his obligations and know his rights. I don't think relations would improve because they must give us our land back. Also, they should treat us well. For example we saw on TV that they accept immigrants from Africa, the Sudan, but they reject Moroccans. This is not the way neighbors should behave. We have many things in common, and many interests in common. We need only some investments, then our country would be similar and even better than Spain, and then Spaniards would immigrate to Morocco. If there is serious investment in those 3,500 km of beaches we have, Spaniards would come to Morocco and look for work. Look, they have stopped when the fishing agreement was cancelled with them. They should respect the norms and the laws when they fish here.

Abdelhak Filali, 42, brass artisan, Fes[36]

What do you think would be best to improve relations between Morocco and Spain?

First of all, Morocco should improve its political and economic situation, to create an industrial opening-up and for the people to have working conditions, so that they won't have to immigrate. And those that are here should be able to find a decent job and get integrated, ... and we could learn from their culture and they from ours.

Blanca Ester Lopez Esparza, 38, teacher, Madrid[37]

What would be the best thing to do that would pave the way for a better relationship between Morocco and Spain?

We have to deal with each other on equal footing. Technological development is not a way or a tool to be used to regard another underdeveloped country as inferior. There should be good cooperation between our two countries. Spanish people should treat our immigrants well. An illegal entry to Spain by any of our people should be dealt with on a sound basis. We want the Spanish to be as fair in their behavior to our people as they are with other people.

Ikbal Al-Hallaoui, 23, economics student, Fes[38]

You told me before that there is a worker on your farm from Morocco. What is your experience with those Moroccans working here in Spain and what do you think about their presence in Spain?

At first I told you that we have bought a farm, we bought it from a client of ours that we have provided the supplies for his farm . . . he had economic problems and said that he wanted to keep on working on the farm . . . so we reached an agreement and he works for us. Then, he said that he promised to be in charge of the farm because we have no time for it, so we reached an agreement. Then, he takes care of the chickens all the year around, but in case he needs someone else to help him with his work, for example, cleaning the place . . . then he calls a man, and it seems that he calls a Moroccan and says that he is the best worker he has ever known. The Moroccan man, he teaches him and . . . but I don't know him, I think that I've seen him once there, but I didn't talk to him, either. Then, he wants us to employ the Moroccan to help him everyday and, if it is economically profitable, we are going to employ him for some time, because he is said to be a very good worker. But I'll also tell you that, not too long ago, they had to go to get the chickens . . . you know that they have to be taken . . . they are grown for two months and then they are taken at night . . . well, he employed some other Moroccans and, once they were there, they blackmailed him, they asked for more money . . . for 2,000 pesetas more for each one of them than had been agreed to . . . I think that they wanted 6,000 pesetas an hour and he had to pay them this money. So, not all the Moroccans are the same. But you asked me what I think about the Moroccans being here, right? It is OK to me. The same as the Spaniards who used to go to Germany, I think it is OK if they always respect our rules.

Juan Luis Martin Rojas, 37, truck driver/owner and farmer, Valdezufre[39]

What would your advice be regarding the improvement of relations between Spain and Morocco?

We would tell them that such a thing is profitable and such a thing is bad. We would tell them that the Spanish are successful; our King visited them. They are successful. We are neighbors and they are the gate to all the other countries. God placed them in a good place; to go abroad, all the people of [Africa] have to cross Spain. We have our brothers, friends, and ambassadors who go to France, Belgium, and Italy. They travel through Spain. God placed them [the Spanish] on the road, and God loves people He places on the road. They are our brothers. We go to them. We eat their food. We spend the night in their places, we have commercial relations with them. We exchange goods and profit. We are neighbors.

Muhammad Chairament, 54, grocer, Ouaoumana[40]

What do you think would be the most important thing to do to improve Morocco's relationship with Spain?

First I would eliminate the borders, the meters and meters and meters of metal fencing; that would be the first thing I would do. When you close something, you are giving it more value. As if that Western lifestyle was worth much more and that it had to be protected. So, if from the outside, you come in and can see if you like that new land or not. I am sure that many who come would go back home if the effort of coming over were not so great. There would be more natural flow. All people have come and gone from one place to another for centuries. That is the natural movement that hinders stagnation. There is a flow of people from one place to another.

Maria José Benitez Rojas, 41, psychologist, Seville[41]

What is the most important thing to do to improve relations between Morocco and Spain?

Everyone should mind their own business. To make relations good, people should manage to forget about the stagnation of the past because if the individual sticks to the historical problems, he will not improve [anything]. The human should try to start a new piece of paper. He should try and overcome the events of the past. Because although the individual focuses on the past, he should instead keep it in his mind but never make it an obstacle to progress. We should move past our history in order to see the future. If we keep our pessimistic view, the view of the colonization of each other, we will not advance and our problems will always exist. We have to get beyond our problems. And I think that this is the only way.

Hichem Si Ahmed Oali, 31, teacher, Fes[42]

Common Ethical Principles: The Lesson of Abraham/Ibrahim and the Strangers and the Conflict Over Immigration Between Spain and Morocco

1. Offering hospitality to strangers is not optional: it is the responsibility of all Christians and Muslims

In listening to the small cross-section of voices presented earlier, it becomes clear that the issue of immigration between Spain and Morocco incites optimism, pessimism, passion, and hope among the populations of both countries. The most constructive advice seems to be to move beyond current perceptions of history, to expand the definition of neighbor, and to place one's own feet in the shoes of the other. This is the path that leads us to question the veracity of what we perceive as the borders that divide Spain and Morocco.

In moving toward a Spain that is more hospitable to its Southern neighbors, it is clear that Moroccan suspicions surrounding the Spanish population's motives and openness to newcomers must be addressed. As witnessed earlier, many Spaniards see in the North African population a group of people not at all unlike themselves: searching for work further north much as the Spaniards did in the 1960s. At the same time, those recent immigrants who have not been able to find work sometimes turn to crime, a fact that has left an indelible mark on many Spaniards' perceptions of North Africans.[43] Somewhat ironically, Spain has recently adopted an immigration measure entitled "a hospitality and responsibility policy," which is enforcing stricter controls on the Spanish borders than ever before.[44] This new policy comes on the heels of a previously more lenient 1999 policy toward illegal immigrants that allowed those who had already been working in Spain for a year to apply for amnesty.[45] Current debates in Spain center on finding a balance between the number of workers needed and the number who enter Spain. Many Spanish interviewees spoke of a desire to only have immigrants enter Spain who have already secured jobs and papers with Spanish employers. Many Moroccan interviewees placed the blame for the huge amount of people fleeing Morocco on Morocco itself, which has been unable to supply living wage employment to a significant amount of the population.

What will it require to make Spain a more hospitable place for Moroccans? What will be necessary to do in order that Moroccans feel safer within Spain? Both countries have mutually beneficial needs: Spain requires a large number of workers to harvest their crops, and Morocco has a high level of unemployment. One might assume that some simple solution should logically present itself. Unfortunately, distrust on both sides has rendered any easy solution temporarily impossible. Spanish farmers are now favoring the use of Latin American and sub-Saharan African workers over North Africans, due to past problems they have experienced with Moroccans and Algerians. This fact, however, has not stemmed the tide of North Africans risking the crossing of the Strait. It has simply changed many peoples' goals regarding the country they ultimately seek to work in.

These observations are based on a series of conversations between the author and farmers who employ foreign workers in Almeria, the region that includes El Ejido. Due to the significant negative publicity surrounding the February 2000 riots in El Ejido, most people contacted refused to participate in a recorded interview, or in many cases even give their names. An exception to this rule was the farmer Juan Carlos Gutierrez. He spoke for many people in his region when he answered the following questions.

What do you think would be the most important thing to do to improve Morocco's relationship with Spain? Imagine that you are the king of Morocco or José Maria Aznar and that you can change whatever you want in order to make these connections better.

Well . . . if we need labor here, I'd make contracts like the ones made for the Spaniards. In Spain we have been emigrants for a long time, many people emigrated to Germany, but they all had their papers and their contract. If it was a three-month campaign, then they spent three months working there and then they came back to their country. So, they should do the same thing here. Do they need labor in agriculture? How many Moroccans do we need? [If] we need 3,000 Moroccans, then make them come here to work, with their papers and their contract. How long do we need them? Three, four, five months? But once they have finished, tell them to go back to their country till the following year. I think that this would solve most of the problems. What happens is that there are many immigrants who come here with no papers, they risk their lives crossing the Strait of Gibraltar and, once they come to Spain, they don't care, they rob people . . . because then they have to wait five or six months for the trial and you cannot leave the country in that period of time. It is a way of staying here, they do something illegal so that they are captured by the police and they say . . . this man is accused of having robbed a shop and the trial will be in eight months time . . . then, they cannot go to any other country and it is the perfect way of staying in Spain.

Javi [the bar owner where this interview took place] told me that farmers here have had so many problems with the Moroccans that now they prefer to hire people from some other places. Is this so?

Yes, that's right. At the beginning, the Moroccans who came here were given a job and a place to live but . . . it's been proven that when you gave a place to live to a couple of Moroccans, in a couple of days there were 8 Moroccans living there, the house was destroyed, dirty, . . . then, after they realized this, they didn't want Moroccans anymore, the farmers rather prefer other kinds of people who understand our customs, people more similar to us.

Which ones are the groups who can assimilate these customs?

People from Argentina, Venezuela, Galicia, Cuba . . . people who can speak Spanish, Catholic people with customs that are similar to ours, people who can understand our way of living . . .

Javi told me that there are people from Senegal who are good workers.

Yes, that's it. Here, there is a group of people from Africa, from Senegal, Ethiopia . . . Black men who are no problem. Once they are done at work they go back to their houses, and there is no problem with them.

What percentage of the workers here have legal papers?

Let's say that . . . for example, if there are 5,000 immigrants, only 2,000 have their papers. The rest of them are illegal immigrants.

Then, most of them are illegal workers.

Yes, most of them are illegal workers. What happens is that now, with all the problems that they have had, the government is trying to get their papers ready.

Is there any police group or someone from the government that comes here looking for the illegal ones?

No . . . they look for them on the beach because they come in a fishing boat . . . they are groups of 80 or 90 people with no papers. Actually, if you were here during the summer, you know that they caught around 600 of them . . . or even more . . . because there must be many of them that weren't caught.

But, once they have a job, nobody does anything about it. They can keep on working with no problem, can't they?

No . . . once they've got a job they look forward [to receiving] their papers in order to go to some other place in Spain to work, to Barcelona or Madrid, for example. They come here looking for Spanish [citizenship] and for their papers so that, once they've got them, they are able to travel around the country looking for another job, because they don't like agriculture. We don't like it and the same happens to them, although they come here with work as an excuse, to [work] in agriculture to become legal in the country and, therefore, to be able to go to Madrid, Barcelona, Valencia . . . to live better.

So, would you say that Las Norias (a part of El Ejido) is a stop in a long way?

Yes, Las Norias, La Mojonera . . . and many of the villages around here are a stopping point on their way in order to get their papers and, once they've got them, they emigrate to some other place.

Juan Carlos Gutierrez, 32, farmer, Las Norias de Daza[46]

It is clear that if substantive hospitality is to return between Spain and Morocco, three things will have to take place. To begin with, Spain must change its immigration policy in a manner that affords more opportunities to Moroccans, while it realistically takes into account the Spanish need for more labor in agriculture and other fields. Second, Morocco must press harder to limit the number of people crossing the Strait illegally. This cannot be accomplished without the introduction of substantive jobs programs, more widely accessible higher education, and a substantial monetary investment by the Moroccan royal family in the lives of ordinary Moroccans. While Spanish investment in Morocco remains significant, it does not always translate into living wage jobs for a large amount of Moroccan citizens. Finally, both Morocco and Spain must draw more deeply on their respective and quite similar traditions of hospitality, rooted in a religious and agricultural culture. There are fewer differences between the generosity of people in the small villages of both Morocco and Spain than either side is often willing to admit. These rural cultures of generosity should be lifted up and celebrated, especially in both countries' larger cities.

2. *It is in encountering the one we call "other" that we recognize the truth about ourselves, and our common Divine origin*

It is not a secret to either the average Spaniard or Moroccan that their common histories have linked them together. They are genetic cousins, whose cultures are historically based on agriculture. They are also two populations that have found varying degrees of economic salvation in emigrating from desperate circumstances to places that offered work and a more dependable way of life. Yet times have changed, at least for the Europeans. Now it is not an uncommon hypocrisy to find Spanish (and even Irish) people complaining about the amount of immigrants in their own countries.

The Spanish, the Irish, and the Americans are not the first people to lose their historical perspective in the face of newfound wealth and opportunity. The question that arises from such circumstances is simply this: what is the best way to refresh a population's memory of its own past reliance upon the help of others? While the details of circumstances vary, the facts do not. Many Spaniards would have fallen into desperate and perhaps irreversible poverty without the opportunity to work in such countries as Germany. The Irish, who at times appear to have dispatched the majority of their population overseas, relied on foreign opportunities to keep financially afloat that were not present at home. Now it is the Moroccans, the Algerians, the Poles, the Ecuadorans, the Ukrainians, the Albanians, and the Senegalese (among others) who are reaching out for the very same kind of help.

Some viable forms of shaping a more positive public opinion on foreign immigrants have emerged from antiracist groups, religious, and secular NGOs, and people whose work promotes cultural understanding across borders. In the city of Seville, there are three groups in particular who are spearheading this work. The first is the Casa de la Memoria de Al-Andaluz, headed by the historian Dr. Sebastian de la Obra. Dr. de la Obra bought a Muslim-built house in the heart of Seville's old Jewish quarter and transformed it into a museum, a concert hall for Andalucian and North African music, and a place that welcomes theater troupes who focus on pointing out Spain's Muslim roots. Another important organization in fomenting better relations between Spain, Morocco, and the world is the Foundation for the Three Cultures of the Mediterranean. An organization whose stated purpose is to promote cooperation among Jews, Muslims, and Christians, The Foundation for the Three Cultures was originally sponsored by King Juan Carlos of Spain and the late King Hassan II of Morocco. The foundation's inauguration was hosted by former Israeli prime minister, Shimon Peres.[47] The foundation exists to hold conferences where Jewish, Christian, and Muslim businessmen and women and scholars can meet and discuss new ideas for cooperation and understanding. Many Moroccans and Spaniards have participated in these conferences. Another important group in Seville is the one that founded the first openly active mosque within the city for what may be centuries. Sufi in its practice, this mosque features many activities for non-Mulims, including a teahouse, Arabic classes, and numerous cultural events. The mosque has attracted many Sevillians, including a growing number of converts to Islam. While their numbers are still small, their presence is critical in reminding Spaniards of the common religious and cultural roots they share with their Moroccan neighbors.

For Spain and Morocco to more readily see their own reflection in the face of the other, three things must take place. First, Spain must not sacrifice its African and Muslim heritage on the altar of the EU. There is no mention in the EU charter of the necessity of forgetting one's history. In fact, nations of the EU are called to preserve, maintain, and celebrate their distinctiveness. Spain without its Muslim heritage is inconceivable. Second, Moroccans must look beyond Spain's current manifestations of colonialism in Morocco and be brought to collectively remember their own occupation of Spain. Morocco was not always a victim; it was once a grand conqueror. Finally, both Spain and Morocco must recognize their countries as one bioregion, through common geographical and biological makeup, shared ocean waters, a common genetic history, and an intermingled cultural reality that defies many efforts at separation.

3. Offering sustenance to strangers can sometimes push us to the edge of our comfort level. What we have not done before is often the most difficult thing to do of all

It is no secret that many Spaniards are afraid of Moroccans. Some of these fears are based on realities, while others are simply based on tradition and ignorance. The same could be said for Moroccans. Truly horrific incidents such as the brutal attack in El Ejido only serve to cement a violent and bigoted image of Spaniards in the eyes of Moroccans. Yet many Moroccans' fears are unfounded. Even among the small cross-section of Spaniards interviewed for this study, the majority expressed positive feelings and even positions of solidarity with Moroccans. A number of the Moroccans' greatest defenders were those who had worked beside Moroccans in Spain. Others expressed how much they loved the land and the people they encountered when they visited Morocco, and how much they were reminded of Spain.

These positive sentiments must be built upon to form policy and institutional decisions. If the Spanish and Moroccan governments are unable to construct more cultural exchanges between their countries, then NGOs must step into the breach. Currently, the vast majority of Moroccan students invited to study in Spain come from the Western Sahara, a policy that is provocative and unproductive in building better ties between all Moroccans and Spaniards. For this reason, more independent groups must become leaders in promoting student exchanges, language classes, and the promotion of a new view of history that focuses on Spain's and Morocco's mutual cultural debts to one another. Every successful promotion of person-to-person contact between these two populations could result in one more Spanish or Moroccan citizen teaching their neighbors the value of learning more about the other nation's people, land, faith, and traditions. It is truly remarkable what one person can accomplish if she or he is given an opportunity.

4. *The world that humans inhabit and the realm of the Divine are constantly intersecting, producing challenges, beauty and fear while defying all expectations. The stranger we encounter could in fact be God or an emissary of God*

In its relationship with Morocco, Spain faces a crucial decision: whether or not it will become a bridge or a wall between Africa and the EU. While many would cite Brussels as the most important player regarding such decisions, it is clear that the EU member states still retain a great deal of leeway in constructing their own foreign policy. The full ramifications of EU membership have not yet been fully defined. For this reason, it is critical that Spain show initiative in tempering Brussels' desire for the Iberian Peninsula to stand as a barrier of protection for a "Fortress Europe." To sharply curtail immigration at this stage in European–Maghrebi relations would be to set the stage for future conflicts that Europe would be wise to avoid.

As the populations in Morocco and the rest of North Africa grow, pressure is building to find solutions to the Maghreb's staggering levels of unemployment. At this juncture, many in Europe are not looking far enough beyond the horizon. It can be argued that current policies better serve to isolate Morocco and North Africa than to integrate them into the social and economic fabric of the EU. Conservative estimates predict an increase in the population of Morocco to four times its 1960-level by the year 2025.[48] Such a spike in the population will potentially dwarf current clashes over illegal immigration. Military analysts at the RAND corporation offer a more ominous prediction: by the year 2006 it is not inconceivable that a North African nation such as Algeria or Libya will have the capability to deliver a nuclear device to southern Europe.[49] For these reasons and many others it is incumbent upon Spain and Morocco to chart a future together that builds North–South relations, and creates a template for other Mediterranean countries to do the same.

If the current basis of Moroccan–Spanish relations is to change, Spain must show Morocco that it has truly broken away from its historical theological claims of exclusivism. Morocco needs to see a new face of Christianity from Spain. Once again, this task can only be accomplished by the efforts of progressive Christian NGOs who are willing to embrace Muslims as Abrahamic brothers and sisters. Such sentiments exist among the Spanish populace, but they remain unknown to many Moroccans. The work of both sides is to arrive at a point that Spaniards and Moroccans can see more clearly the interdependence of their faiths, cultures, and ecological futures.

*5. Extending help to strangers is a concrete expression of practical faith and
solidarity, for today's host does not know when he or she will be forced
to rely on the compassion of strangers tomorrow. Conversely, those who are
recipients of aid must also be people of faith and be open to receiving what they
are being offered*

Sustainable Diplomacy is systemic diplomacy. It relies upon the efforts of multiple governmental, NGOs, and individuals to simultaneously strive toward building connections, dialogues, and new understandings between populations, and not simply heads of state. As a single, contiguous bioregion, Spain and Morocco must cultivate a mutual respect that makes space for cooperation on the joint management of natural resources, expansion of the sharing of ecologically appropriate high technology, and the increase of the exchange of students, scientists, artists, farmers, and other ordinary people. Together, Spain and Morocco must acknowledge the truths presented by the Ecological Footprint and base their local, regional, and national consumption and waste management policies on their true biological capacities.

Finally, Spain and Morocco must rise to the occasion and begin to see their full potential as partners who could one day serve as a new model of cooperation between Muslims and Christians. This new relationship would be based on the mutual recognition of living on a common land, inhabited by a common people who share a common history. The similarities, ties, and common aspirations identified by the interviewees cited are evidence that such a relationship is possible. It is therefore for the two populations themselves to explore various means of building ties when government-level dialogue breaks down. The extraordinary creativity that exists within both populations makes such a proposition possible.

The facts presented by Ecological Realism make the building of such ties a requirement for long-term survival.

Points of Disagreement: The Lessons of Abraham/Ibrahim in the Biblical and Qur'anic Accounts

1. For the Christians, Abraham was the first Jew, and for the Muslims he was the first Muslim

Spain and Morocco's relationship is impeded by competing claims of identity. While Spain views itself as the arbiter of how the relationship will or will not progress, Morocco's leadership sees its country as a natural addition to the EU's continued expansion. While it appears clear as to who is the conquered and who is the conqueror, there is now a new type of influx (which many Spaniards see as an invasion) from the South taking place, in wave after wave of small boats washing up on Spanish shores.

Spain, and the EU as a whole, seems to have determined that Morocco's progress will be realized through a mimicking of the Northern neoliberal economic model. In this vein, the EU signed an agreement with Morocco to create a free trade zone between their two "entities" that will take effect in 2012. King Mohammed VI responded to this arrangement by stating "you cannot make the movement of goods free but close it to people."[50] Spain, with strong encouragement from the EU, has thus named itself as gatekeeper, while naming those who live south of the Strait as "other." Despite these divisive distinctions, Spanish and Moroccan origins and realities continue to hold many things in common. Geographically, geologically, genetically, historically, and biologically, their origins are the same. Disputes to the contrary accomplish no more than the competing claims over "ownership" of Abraham. There is only one Abraham. There is only one bioregion, and it must be shared between the citizens of Spain and Morocco. How this will be accomplished is the work of Sustainable Diplomacy.

2. Many Christians and Muslims are equally wedded to the notion that their faith is the exclusive source of salvation in the world

Spain views itself as a major player in Morocco's economic salvation. Spanish investment in Morocco, while currently more modest than in past years, remains robust. While it is clear that many of Morocco's problems can be attributed to economic shortfalls, it remains the source of rich resources and potential on its own. Yet it is the Northern model of development that is all but assumed by Moroccans to be the template for meaningful progress.

To what degree is the neoliberal model of economic development salvific? What would be gained and what would be lost if Morocco were able to realize all of the material advantages of its northern neighbor? Is there room in this model of development for all of the gifts Morocco already possesses? The thriftiness, tenacity, and creativity of the Moroccan people are assets that cannot be expressed in a quarterly earnings report. The spiritual depth and knowledge of the land that many average Moroccans possess could well be lost in the arrival of development as it is envisioned in the North. Morocco needs a new definition of development that emerges from its own experience in a postcolonial world. This new arrangement should not be based on an employer–employee model, but built on the basis of a real solidarity with its Northern neighbors. Plummeting European birth rates suggest a future of mutual benefit across the Strait. As Migration News reported:

> Europe is aging. By 2025, an estimated 113 million
> Europeans—nearly one-third of the population—will
> be pensioners. Europeans retire on average at age 61 . . .
> According to U. N. projections, the 15-nation E. U.
> would have to accept 47.4 million immigrants by 2050
> to keep its population at the current level of 372 million.[51]

It is clear that Europe will eventually be compelled to realize that it needs North Africa as much as North Africa needs Europe. Political and economic salvation thus lies in mutual cooperation and a deliberate end to many types of exclusivity. How slowly or quickly the populations of Spain and Morocco come to realize this truth will depend in great part on the efforts of the practitioners of Sustainable Diplomacy.

CHAPTER 6

THE FUTURE OF SUSTAINABLE DIPLOMACY

Er querre es cuesta arriba	To love is all uphill
y el olvidar, cuesta abajo;	And to forget is all downhill;
quiero subir cuesta arriba	I want to go uphill
aunque me cueste trabajo.	Although it will cost me dearly.

    —*soleá*, traditional[1]

Miralo por onde viene	Look at him as he comes
agobiao por el doló	Bent double with the pain
chorreando por las seines	Dripping from his temples
gotas de sangre y sour	Drops of blood and sweat.
Y su mare de penita	And his mother in anguish
destrosao er corazón	Her heart broken.

    —*saeta*, traditional[2]

Abrase la tierra	Let the Earth open
Que no quió bibi!	'Cause I don't want to live
Si mas en er mundo	If in this world I never hear
la bos e mi hermano	my brother's voice again.
No la güerbo a oír	

    —*siguiriya*, traditional[3]

The conduct of Sustainable Diplomacy is founded on an intimate understanding of the relationship a population has with its land, its religious and cultural traditions, its ecological reality, and its neighbors. These five factors encompass a wide spectrum of information and cover a great many details. Much of this information cannot be gleaned from books; it is acquired through conversation, observation, and participation in ritual practices with a willingness to authentically enter into people's lives. We can no longer rely on the information supplied by a few influential people residing in capital cities in order to make policy decisions. We must look for direction in villages, across the economic, ethnic, religious, geographic, gender, and racial spectrums. Often, our first clear understanding of a people comes out of their folkloric traditions. Such traditions are rich with history as it is told from the underside, reflecting the understandings, moods, and outlooks of many ordinary and extraordinary people.

Flamenco music is defined by its rhythms and intended uses. While many in the twenty-first century simply equate Flamenco with a music of dance, its roots are far more profound. Brought to Spain by the Roma, who traveled through eastern Europe and northern Africa, Flamenco is rooted in the experience of the most marginalized peoples of Spain, who melded their musical traditions with those they encountered in their travels. Muslim culture is among many that have affected what is considered to be the quintessential musical expression of Spanish culture. Roma culture is one of the many vital links that bridge Spanish and Moroccan culture. For anyone who has spent time in both countries, the parallels are clear. Flamenco is embraced by most Spaniards, European, and Roma, and is considered to be among Spain's most unique contributions to global artistic culture. The words of Flamenco can often stand as expressions of the voices of many ordinary Spaniards, with its remembrances of joy, loss, hope, and despair. It is a living tradition that bridges cultures.

Flamenco is traditionally sung as a four line *copla*, repeated in different ways throughout the song. The first passage above is a *soleá*, and is meant to be sung without accompaniment. It is one of the root forms of true flamenco, and its mood can range from "tragic to frivolous."[4] In the soleá presented earlier, the theme of the difficulty of love is featured. Love is an "uphill" proposition. Even at a high price, it remains a proposition worth pursuing. The same could easily be said regarding the improvement of relations between the Spanish and Moroccan populations. True reconciliation is never easy. It is an uphill battle that must be fought on a daily basis. Yet it is a battle worth waging against racial prejudice, economic oppression, religious exclusivity, and ecological destruction. Love of the land and all of its people must now become the task taken up by both countries to ensure their long-term mutual survival.

The second song presented is a *sieta*, a song that is often sung during Holy Week during the procession of *pasos*, large floats that are carried through the streets and present one of two figures. The first paso in the procession often depicts a particular moment in the life of Jesus from the time he makes his final entrance into Jerusalem until his resurrection. This paso of Jesus is then followed by a paso carrying an image of Mary weeping, her arms held out as if she is carrying a baby who is no longer there. A *sieta* is sung acapella at particular junctures in the procession, most often to Mary, but sometimes to Jesus. As the procession stops and everyone falls silent, a singer often emerges on a balcony above the crowd to sing a haunting and passionate lament. Among the many things that are striking about the sieta is that its melody and cadence often mimic the call to prayer issued from a mosque. In the sieta presented, the lament is for a woman who must watch her son die a brutal and unjust death at the hands of a colonial power. The song could just as easily be sung by a Moroccan woman whose son was lost at sea trying to reach Europe. Then again, such a story speaks to the struggles of the Spanish people, who were until very recently less than a "first-world" nation, and whose young people struggled under foreign employ. This is one of many instances where a religious story tells the story of those who relate it. It is a story of mothers who have cried and sons and daughters who have suffered and died. In many flamenco lyrics, the distinction between the Virgin and one's own mother is blurred. Both are loved, both lament a loss, and both elicit a passion that is without parallel.

The final *copla* is taken from a *siguiriya*, often one of the most tragic of all flamenco songs. The *siguiriya* presented earlier is clearly mourning the death of the author's brother. Flamenco is highly personal. Yet flamenco makes the private public. Sustainable Diplomacy must echo flamenco in this regard. It must draw from personal and sometimes private realities the truths that will change public policy. Distinctions appear to be widening between those called brother or sister and those called "other." The time has come to fully lament the fact that many Moroccans and Spaniards fail to see that they are brothers and sisters, despite the fact that they are only separated by a small stretch of water. "Let the Earth open" sings the *cantador*. Yes, let the Earth open, but not to swallow the mournful, but rather to open their eyes to the reality of a fuller life. A life of connections made; a life of solidarity. Let the Moroccan and let the Spaniard see again their connections, their interdependencies, and the common hopes and fears. Let Spaniards and Moroccans embrace the full meaning of the word "respect"—one that is mutual, and one that looks again with new eyes upon problems long unresolved.

The Common Ethical Principles: A Compendium

In the course of our examination of Moroccan–Spanish relations, we have identified 14 common ethical principles that Christianity and Islam hold, which have the potential to promote the resolution of political conflict between Morocco and Spain. These principles are:

1. The human, the Earth, and all of its creatures are in a common, inter-related ontological relationship.
2. The human is in an intimate, evolving, and dialogical relationship with the Creator God. This relationship is shaped by human action and the human's ability to understand the meaning of responsibility to greater Creation.
3. The human is in an intimate and evolving dialogical relationship with human and nonhuman Creation.
4. The human is in possession of tremendous power vis-à-vis Creation, bestowed by God's very design.
5. There is a language within and expressed by Divine Creation.
6. We live precariously and temporarily between heaven and hell, both of which can be reflected in earthly life depending on the path of human action.
7. There is an *axis mundi* in God's Creation, providing balance, wisdom, cohesion, and logic to the biosphere. It can be seen and understood by those who look for it and strive to interpret it.

8. Creation was designed to support, feed, and heal all of its inhabitants. When Creation fails to do this, it is due to human selfishness, jealousy, and sinfulness.

9. Two of the most followed figures in human history, Muhammad and Jesus, emerged from a Creation that they were called by God to exegete for their adherents.

10. Offering hospitality to strangers is not optional: it is the responsibility of all Christians and Muslims.

11. It is in encountering the one we call "other" that we recognize the truth about ourselves, and our common Divine origin.

12. Offering sustenance to strangers can sometimes push us to the edge of our comfort level. What we have not done before is often the most difficult thing to do of all.

13. The world that humans inhabit and the realm of the Divine are constantly intersecting, producing challenges, beauty and fear while defying all expectations. The stranger we encounter could in fact be God or an emissary of God.

14. Extending help to strangers is a concrete expression of practical faith and solidarity, for today's host does not know when he or she will be forced to rely on the compassion of strangers tomorrow. Conversely, those who are recipients of aid must also be people of faith and be open to receiving what they are being offered.

These ethical principles are hardly the only ones Christianity and Islam share. There are many more to identify and use in the course of promoting conflict resolution. Determining how these principles are to be used is the task of future practitioners of Sustainable Diplomacy. This much is clear: the future of Spanish–Moroccan relations demands a different approach, for as we shall see, the future of this relationship will set the tone for many other larger arenas of cooperation and conflict.

The Future of Spanish–Moroccan Relations

In the context of EU–Maghrebi relations, the Spanish–Moroccan relationship is pivotal. While Spain has been named as the EU's first line of defense against illegal immigration from the South, Morocco stands as the last stage of a long journey many northern and sub-Saharan Africans have made en route to attempting a crossing into Europe. Together, Morocco and Spain comprise one of the most well-traveled geographic bridges in the world. It is for this reason that the future of a large portion of human migration into Europe is dependent upon closer ties and more coordinated cooperation between Spain and Morocco.

However, Spain and Morocco are not simply a bridge between two economies; they stand together as a repository of knowledge of an intimate interchange between Africa and Europe that can be matched by few if any other nations. For this reason, it is incumbent upon Spain and

Morocco to work harder to resolve their differences, so that they may one day be a model of cooperation for other European and African nations. Therefore Spain and Morocco must strive to define what it means for two such nations to stand on more equal footing, as they promote mutual respect along with a higher level of political, cultural, commercial, and ecological cooperation and exchange.

In the context of the North–South split, Morocco and Spain are among the world's most proximate players. Each country is defined by its continental neighbors as a nation strongly impacted by its geographic nearness to another continent. Morocco and Spain are in a position to build bridges that few other countries could ever imagine. Circumstances such as Europe's low birthrate and North Africa's growing population necessitate new and bolder guidelines for future human migration. At the same time, Morocco stands as one of Spain's best opportunities to retain its identity within the context of the EU. This is because Morocco is increasingly the repository of sustainable knowledge that Spain is daily losing in its effort to distance itself economically and culturally from the Two-Thirds World. For these reasons, Spain and Morocco have an opportunity to redefine the North–South split from being a theater of Northern patronage to a place of real two-way exchange. Both could benefit from this type of relationship. It is for the people of Spain and Morocco to recognize this fact and act upon it.

In the context of Muslim–Christian relations, Spain and Morocco could one day be leaders. The Spanish–Moroccan relationship has many advantages not enjoyed by other nations. Each has conquered the other. Both share a common genetic, geographic, and biological reality. It is therefore up to the more progressive voices within each respective religious tradition to use these commonalities as concrete building blocks to a deeper level of mutual understanding and cooperation. The common ethical principles presented in this book are only a beginning; there is so much more to build upon.

The Goals of Sustainable Diplomacy

Spain and Morocco must work together *to build on the truth of Ecological Realism*. Only through seeing themselves as a common bioregion rather than two separate nation-states will Morocco and Spain become ecologically sustainable. Both nations must concede the commonalities of their respective Ecological Locations, and the subsequent common histories their human and nonhuman populations share as a result. Looking at the role the land has played in the lives of both populations will provide multiple opportunities to use religious traditions as the basis for cooperation rather than conflict. Growing resource scarcity only heightens the necessity of pursuing this direction. The relationship Spain is now forging with the EU is new and in many ways uncharted. Spain's multiple proximities to Morocco necessitate that it remain open to deepening its partnership with

Morocco even as it becomes more integrated into the EU. While the foreign policy the EU is officially produced in Brussels, each member state has retained significant latitude to chart its own diplomatic course. The Ecological Footprint will be a critical tool in forming policy for both Spain and Morocco. Such an approach must be embraced as a common currency of survival and sustainability, just as the Euro has come to supplant the traditional European monetary system.

Spain and Morocco must learn *what it means to protect existing sustainable communities while building new ones*. As Larry Rasmussen has written, "Earth's treasures are not an unlimited resource but a one-time endowment, essentially a closed system that must sustain itself or die."[5] Ecological Realism demands that Morocco and Spain make this realization. This means that both Spain and Morocco must first move to support sustainable communities that already exist, holding them up as models for others and building on their wisdom. As Rasmussen has written, "the Northern concept of 'progress' has failed."[6] Therefore, Morocco's or Spain's economic salvation does not lie in mimicking the highly consumptive, sustainability-stealing models presented by many Northern nations. Morocco and Spain must learn what it means to be sustainable in their own bioregional context, letting the bio-capacity of their land and ocean shares determine their mutual level of consumption. "Justice and equality are central to establishing and maintaining sustainability."[7] For this reason, Spain and Morocco must make the connection between ecological sustainability and human rights. Both nations have been cited by Amnesty International for their poor records regarding terms of imprisonment and even torture.[8] Correcting these patterns are only the first steps in promoting truly sustainable communities in Morocco and Spain. They are steps that must be taken sooner rather than later.

The normative guideposts of Sustainable Diplomacy are many. As mentioned in chapter 1, the principal normative guideposts or framework of Sustainable Diplomacy include the following: solidarity, participation, sufficiency, equity, accountability, material simplicity, spiritual richness, responsibility, and subsidiarity.[9] Upon further examination, new guideposts have emerged and will continue to emerge. Among the new guideposts that have emerged from this text are mutuality, intimacy, spiritual and material transformation, hospitality, respect,[10] the embracing of the "other," and an absence of religious exclusivity. In time, more normative guideposts will be added to Sustainable Diplomacy's framework. Some will be bioregionally specific, while others will be applicable across the spectrum. Building on this list of markers will be the work of future practitioners of Sustainable Diplomacy.

If Spain and Morocco are to become sustainable, they must do so through *building a common bioregional identity*. This common identity can provide the context for the freer flow of humans, knowledge, and goods. A common bioregion also broadens the scale upon which one can draw conclusions regarding the Ecological Footprint's definition of "Fair

Earthshares." Were Spain and Morocco to share together the responsibility of becoming sustainable for the long run, their chances for success would be greater. At the same time, traditional definitions of individual state-to-state diplomacy would be subsumed by a more systemic, multilevel manner of building ties. A bioregional approach opens more room for the work of secular and religious organizations and communities, including the NGOs, to broaden the Moroccan–Spanish dialogue to include voices that have never before been heard at the policy making level. Such dialogues would not be confined by the dictums of state-to-state negotiation approaches, but instead they would work beside existing channels to create a deeper, more multifaceted dialogue that better represents the entire population of each respective country. This is the task of the practitioners of Sustainable Diplomacy, as informed by the goals of Ecological Realism.

The Impediments of Implementing Sustainable Diplomacy

Moving from an anthropocentric, secularized state-to-state diplomacy to the practice of a truly Sustainable Diplomacy will prove extraordinarily difficult for a number of reasons. To begin with, one can never underestimate *the "cult" of the nation-state* and its power to endure in the minds of many populations. Nationalism is a card that is often played by world leaders in an effort to conceal systemic inadequacies from the populations they represent. Personal identities are often tied to artificially created national identity, which may serve the purposes of a minority of people over that of the majority of the population. Nationalism and national identities are powerful tools that will not disappear in the short term, yet their long-term erosion can likely be engineered by those who are consistent in their advocacy of an eco-centric worldview. As resource scarcity increases, there will be two clear paths to follow, and it will be to the practitioners of Sustainable Diplomacy to make as many arguments as possible for turning away from nationalism and turning toward the Earth.

The power of globalization is yet another potent phenomenon that undermines the construction of a Sustainable Diplomacy. In many respects, as globalization extends itself, our monetary systems reflect less and less of the reality that the Earth has a limited capacity to sustain the global biosystem. It is not the interconnection of economic systems but the scale and depth of globalization that undermines sustainability. The critical notion of subsidiarity—a bulwark of sustainability—is lost when we become dependent upon fruits, vegetables, or other products that are grown or manufactured thousands of miles away. Sustainability requires us once again to value local consumption of locally grown and manufactured products whenever possible. At the very least, the petroleum currently required to move goods and services on a globalized scale is a short-term commodity, one that could be exhausted within decades. At the same time, the pollution generated by such far-flung exchanges of goods and services is making a sizable contribution to the destruction of biosystems on which all Earth's inhabitants depend for survival. It is true that the systems of communication pioneered by the architects of globalization could be used to promote the undermining of many destructive practices. The experience of the Zapatistas in Chiapas, Mexico demonstrated the truth of this. If the power of private economic interests to supplant the importance of democratically elected and nondemocratically elected governments continues to grow, the practitioners of Sustainable Diplomacy must be all the more creative in their methods and approaches in combating these trends. Strategies must be formed that can respond to actions taken by any multinational economic actor, in spite of its ability to rapidly change its identity and location.

Sustainable Diplomacy is also in many ways undermined by *the growing phenomenon of "tribalism"* among many of the world's peoples. It is through tribalism that nation-states are broken down into even smaller pieces, while long-held ethnic resentments, and in many cases outright conflicts, are given free space to flourish.[11] Many such movements express a very clear self-understanding of their own Eco-Location coupled with a desire to correct past and present injustices, but they often do so in an exclusivist manner. Ecological Realism builds community by integrating individual ethnic, religious, and cultural identities in a way that reveals the commonalities that many groups share with other peoples in their respective bioregions. Ecological Realism does not demand that anyone abandon their individual or group identities. For this reason, Ecological Location is a critical component to assess situations on the ground, and invites people to value and to tell their stories while building on existing sustainable communities that have managed to cross anthropocentric boundaries.[12]

The lack of basic human rights in many nations, including Spain and Morocco, diminishes the capacity of any population to conduct Sustainable Diplomacy. While Amnesty International has recently taken Spain to task for its often brutal treatment of immigrants and other non-Europeans, Morocco is also a nation that still lacks many of the institutional components required for democracy.[13] Ecological and human justice are intrinsically and inseparably linked. Ecological Realism embraces the requirements for supporting truly sustainable communities, ones that reject the growing gulf between the wealthy and the poor, the gross inequalities between people of different races, religions, ethnicities, and genders. As Rasmussen has noted, "any power that does not go to the places where community and Creation are most obviously ruptured and ruined is no power for healing at all."[14] Contributing to the support of sustainable communities requires that we "regard all people not simply as individuals, but as persons-in-community who require a living wage, real health care, a clean environment and consumer protections."[15] For this reason, practicing Sustainable Diplomacy means rejecting ecological racism in all its forms, and imploring all communities to share equally the burden of human pollution, along with the fruits of the biosphere.

Ultimately, one of Sustainable Diplomacy's greatest opponents is *our increasing collective inability to imagine a better, more egalitarian, and cooperative international community*. Many people have lost faith in both international institutions and themselves to promote meaningful international cooperation. This problem can be attributed to a number of factors, including a growing number of international conflicts and the erosion of the power of traditional institutions such as the United Nations to mediate conflict. Current trends in U.S. foreign policy value unilateral action over transnational cooperation. Simultaneously, globalization's fruits are unequally shared and have all but bypassed many two-thirds world nations, particularly in sub-Saharan Africa. These trends have been exacerbated by many in the North who seek separate means of survival, while those in the South find themselves increasingly at the mercy of the fluctuations of the North's economic agendas. In the face of all these developments, the challenge to have faith in the possibility of increased and nonexploitative North–South cooperation is daunting. For this reason it is critical that we find new sources of faith, or rediscover and reinterpret our faiths of old.

Fighting Against the Cross-Tradition Sins

Among our many challenges on the road to sustainability for both Christians and Muslims is to face squarely our enormous individual and collective capacity for ecological *greed*. This is particularly true in the North, but is also a problem among those in the South who live in the same manner as wealthy Northerners. Many unsustainable practices are repeated again and again despite an intricate and intimate knowledge of their ultimate result. From Spain's fishing practices to the unwillingness of the United States to ratify the Kyoto Protocol on climate change, we live in an era where many have lost any sense of personal or collective shame. On the contrary, it could be said that many nations, as well as individuals, take great pride in their ability to consume natural resources well beyond their Fair Earthshare. These same parties are often the first to recommend the same practices to the so-called developing nations as the clear and uncontested road to economic success and personal and collective fulfillment. Sustainable Diplomacy, guided by the Ecological Footprint and the religious traditions of the nations it engages, must strive to rouse within the most gluttonous populations a sense of responsibility for their actions and an awareness of their ultimate connection to nations and individuals they have never personally met. This is a daunting task, but one that must be taken up with vigor and creativity. Respect for Creation and the advantages of such respect can be argued in both religious and secular terms. Sustainable diplomats must be fluent in both languages if they are to be successful.

While it is incumbent upon many to confront their own greed, equally debilitating sins against sustainability are found in *shortsightedness, willful blindness, and calculated denial*. One of the many effects of globalization on the world economy has been to link individual local labor practices, production goals, and consumption patterns to shorter and shorter regional, national, and international business cycles. Where 20 years ago many would have considered true investment to be a long-term enterprise spanning years, the Northern drive to link production patterns to quarterly earnings reports has served as a disincentive to sustainability. For example, under such circumstances the clear cutting of trees often appears to make more economic "logic" to the timber industry than a calculated long-term selective cut. Fruits and vegetable harvests that were once tied to seasons are now often produced year around by high-intensity nonsustainable farming practices, which produce yellow and red peppers in winter along with exhausted soil and diminished aquifers. What is economically expedient is not necessarily ecologically intelligent; these are the lessons of both the Ecological Footprint and Ecological Realism. Central to maintaining such unsustainable practices is a willful blindness to facts that are now a matter of public record. Thus, while many in the economic North are working in anticipation of substantial new car sales in China, we are simultaneously confronted by increasing signs of climate change due in part to rising carbon monoxide emissions. On a grander scale, the world population's migration to urban areas creates a new type of unsustainability: the mega-city. One need not go beyond the borders of the United States to view this phenomenon. Los Angeles has not been self-sustaining for years. It requires water and electricity from nearly half way across the United States to meet its growing needs. It is only through a collective, calculated denial that any country continues to support the growth of such urban centers. Learning from the biosphere itself is to practice a type of balance that precludes our current trends toward higher and higher levels of urbanization. Ecological Realism requires that we see our actions through to their physical end result, and not simply in terms of the next quarterly report.

Just as greed and short-sightedness can undermine sustainability, so too can *egoism*. While believing that one's own community, region or nation is special is not in and of itself harmful, stealing the sustainability of others to maintain one's own high levels of consumption is unacceptable. Egoism in this regard is anticommunity and denies the already existing ties that nations have made across bioregions. A new type of tangible solidarity must be promoted whereby nations with large Ecological Footprints can more easily come to place their feet in the shoes of others, and see the connection between their own unnecessarily high levels of consumption and the ecological and economic poverty of others. Our current state of affairs calls for a new level of ecological literacy across the board, where the connections between life in the Two-Thirds World and the so-called Developed World is more compellingly presented. People in the North and those who live a Northern lifestyle in the South must see that their own survival is intrinsically connected to the survival of the more marginalized. This task can be accomplished in the presence or absence of egoism. This is because this task is most readily accomplished through a calculated promotion of fear, which can trigger the survival instincts of even the most gluttonous consumer. Those who promote Sustainable Diplomacy must therefore not be afraid to act pragmatically. There are too many people who must be awakened to the realities they will soon face, and there is little time to always be polite.

To honor Creation is to honor one's self along with all that is encompassed in the biosphere. *The unwillingness to honor Creation* is one of the principal impediments to achieving sustainability. Honoring Creation means respecting all of one's neighbors, not simply those whom one likes. Honoring Creation, therefore, includes a complete rejection of ecological racism, the practice of dumping waste or housing hazardous materials in the neighborhoods and lands of the marginalized. It also means that it is completely unacceptable to force any human being to work in environmentally hazardous conditions from which those in the dominant racial, ethnic or religious group are exempted. Honoring Creation means living within the bio-capacity of one's own region and freely sharing with others who cannot afford the technology that will make it possible for them to do the same. Honoring Creation also means honoring nonhuman Creation, which is most often without a voice in decision-making processes. Honoring Creation means modeling human lifestyles in an Earth-centric manner, mimicking in style, consumption levels and practices the inherent balance that is contained in nonhuman Creation. Finally, honoring Creation demands that the North abandon the egoism that prevents it from seeing peoples in the South—particularly indigenous peoples—as sources of great wisdom regarding how sustainable communities have been made a reality.

If the North is to honor Creation as it has been described, *many people must abandon their unproven faith that either the market or technology will rescue humanity* from the results of its collective excesses, denial, greed, or short-sightedness. Technology and the market have created extraordinary things. To depose the false gods of technology and the market is not to abandon them in a Luddite manner. Rather, the responsibility to locate, promote, and learn from existing sustainable communities requires an openness to a new relationship with other communities, along with a clearer understanding of the uses and abuses of technology and the market. Technology will clearly play a major role in building sustainability. The manner in which technology is disseminated, however, cannot be based solely on the economical profit motive, but must include an ecological profit motive as well. Northern nations must come to see that it is in their own self-interest to share many of their most sophisticated existing technologies with nations in the Two-Thirds World, so that they might avoid the most ecologically damaging stages of industrialization. This will require taking a chance on a different type of valuation. Diminishing one's own or someone else's Ecological Footprint yields a tangible value to all countries that money cannot match. Artificial capital must be seen as having less value than Earth capital. Current economic means of valuation are ethically and ecologically inadequate. The value of Earth capital is beginning to be recognized and accepted. In the future, the recognition of the value of Earth capital will quite possibly increase and deepen. The problem we will most likely face will be in terms of equitable accessibility to resources.

The Cross-Traditional Ethical Principles

We have already named the common ethical principles held by Islam and Christianity regarding Adam, the Two Trees, and Abraham and the strangers, all of which can be used in promoting conflict resolution. Now we draw from these specific stories some general principles that can be applied under all circumstances and in all efforts to support a Sustainable approach to Diplomacy. These principles include the following mandates.

We must strive to see and to know all of our real neighbors　The practice of living in a truly inclusive manner is something that eludes the average human. Both Christianity and Islam were affected by the Greek proposition that the body and the soul were separate entities. In like manner, followers of Christianity and Islam have adopted anthropocentric worldviews and anthropocentric cosmologies that separate humankind from the planet. Turning to Earth requires a turning away from the arrogant assumption that the human is the crown of Creation. This is a difficult proposition for both Christians and Muslims. For too long, the dominant interpretations of both faiths and their Creation stories have placed humanity at the center of all, and relegated greater Creation (including countless marginalized humans) to the status of material to be used for human gain. This is an ontological lie, and it must be unmasked as quickly as possible if humankind is to survive. Our neighbors are *all* humans, for all humans are created by God. Our neighbors are also found in the Divinely created nonhuman world, which teems with life whose true value we barely recognize. "The well-being of Earth is primary. Human well-being is derivative," writes Brian Swimme and Thomas Berry.[16] This is one of the most difficult lessons of all for humans to grasp. Egoism, greed, narcissism, and denial prevent this truth from being realized. Such sins are committed by Christians and Muslims alike. Now is the time to seek forgiveness and change our collective behaviors. Our individual motivations for change are less important than the collective change itself. This is because our time to change is regretfully limited. Practitioners of Sustainable Diplomacy must be teachers of true sustainability, a sustainability modeled on Earth's own patterns and capacities. Appeals to religious norms can be highly useful in this undertaking, for religious norms offer language familiar to the cultures that produced them. The simple human desire to survive will also be very useful as well.

We must share equally with all of our neighbors   Such a goal cannot only be accomplished through isolated individual action, though such action should not be undervalued. Rather, we must strive to work together in existing communities as teams whose goal is to live within the current Fair Earthshare of 2.1 ha/cap. Such a goal will require a coordinated response that will begin in small communities and move outward. Living in this manner must be presented as living out a moral norm, one that shows tangible concern for future generations, for the generations to come must always be numbered among our neighbors. For this reason, the three following acts should be understood as sin:

1. The refusal to act as responsible representatives of God [*imago dei/khalifatullah*] who value the lives and well-being of all other members of Creation.
2. The injustice of grabbing more than one's due.
3. The arrogance of treating the Earth and its inhabitants as property at one's disposal.[17]

Sharing equally with our present and future neighbors will be particularly hard in the economic North, where immediate, individual privilege is often valued over long-term collective well-being. We must therefore try to concretely visualize a day when someone would be ashamed to be seen driving an SUV in public. Public policy is worthless if it does not filter down to individual lifestyles. Who then will be the models of this new sustainability? Who will teach people how to live a different way? Who will help people make the connections between their current lifestyles and the possibilities that will be closed to future generations if more people do not change? One source of leadership must come from religious institutions that already exist. Within the bodies of institutions that are committed to sustainability, leaders must emerge who are willing to speak out and to admit that past interpretations of religious tradition and Scripture have helped to create the problem we now must fight. In many respects, this will require significant reform of many political and religious institutions. We need new Earth-centric interpreters of the Bible and the Qur'an, who are interested and capable of communicating with broad cross-sections of the population. Muslim and Christian leaders in both the political and religious sectors must be open to learning from other faiths as well as their secular brothers and sisters, and to valuing their insights and advice.

In addition, many in the North and in the South would do well to heed the wisdom of the indigenous peoples of their bioregion, for they are often the most significant repositories of wisdom capable of very clearly mapping out the contours of sustainability. In Spain, this could mean looking to the Basques in a new and more positive light. In Morocco, honoring the indigenous and asking for help would mean to redefine Arab Morocco's relationship with Morocco's Imazighn community. Now is the time to learn how to ask indigenous peoples, among the most marginalized peoples in the world, for help. This will mean different things in different regions. One thing is true—this type of change will require many things from those in power: the ability to admit past and current transgressions against indigenous peoples, the willingness to offer sincere contrition and reparations, and the openness to learn from those who have been made to be the human "throw-aways" of "modern" society. Everything depends on how seriously we intend to become collectively sustainable in our practices. Sharing must no longer be limited to an "inner-tribal" practice. It must recognize the reality of bioregions and the connections that exist among all the people who live in them. Ultimately, we must grapple with the necessity for bioregions to cooperate on an even broader, global level. We are connected to people we have never even met, many of whom we have treated horrifically. We will never grasp the true meaning of "neighbor" until we concede this fact.

To truly share with our neighbors is to share everything. Such sharing includes recognizing that "healing begins in mercy, where God's own strength is tapped . . . Sustainable communities . . . are therefore found in entering into the predicaments of those who suffer, for compassion (suffering with) is the passion of life itself."[18] This is a powerful definition of true solidarity. Those who live in the economic North must learn that writing a check is not enough. We must put our bodies on the line in the fight to realize a truly lasting form of sustainable existence. We must learn from existing communities who practice such solidarity. Until humans, particularly in the North, are able to see that their survival is dependent upon the very people they have marginalized, long-term sustainability will remain nothing but a dream.

In fighting for sustainability, we must honor each other's lands and faith traditions
This means working toward an end to exclusionary claims to salvation on the part of one faith tradition over all others. Religious tolerance is not enough to secure sustainability. Nothing less than real inclusion and full respect will suffice. This means that members of all religious traditions must make a public confession that no one faith has cornered the market on the truth concerning God, humanity, or greater Creation. Instead, we must come to acknowledge that each faith holds sacred truths that deserve honor, respect, and study. No one can underestimate the difficulty of making such a confession for any member of any faith tradition. Yet our fears cannot prevent us from moving in new directions. Honoring the land of another people is a good place to begin such work. Honoring land is a tangible good that accomplishes more than any set of words ever will. By not stealing your neighbor's land, by not depleting their natural resources, and by honoring the people of another nation-state, you are honoring the land. These are among the first steps toward living in a common bioregion. Those religious and secular individuals who practice these disciplines must be lifted up in the construction of public policy, in the reformation of religious institutions, and in the gathering of persuasive political and

economic power potent enough to inspire the even most egregious ecological transgressors to change their practices.

The work ahead is enormous. Considering its scope and its true ramifications can be overwhelming. Yet the work of supporting sustainable communities must be based on the above principles as we struggle to *build bulwarks against the excesses of globalization*. The increasing dominance of the neoliberal model of economics need not be a phenomenon that controls all areas of the globe. Creative efforts by individual communities, such as the ones in Chiapas, Mexico have shown that people can successfully fight against being "developed to death."[19] How people choose to work to reclaim the right to produce their own goods in their own regions must be taken on in a case-by-case manner. Some very simple approaches must be revisited, such as planting whenever possible one's own vegetable garden, or planting a garden together as a community. People need to spend more time talking to the elders of their communities, and learning how they provided for themselves before it was possible to purchase all of the goods that are currently on the market. At the same time, working to provide for one's self is not a call for xenophobic economic policies; rather it is a call to diminish our consumption of fossil fuels by seeing what can be produced in one's own bioregion, while working to form stronger trade alliances with communities in the closest proximity to one's own. For this reason, room must ultimately be made for systemic responses—ones that reflect the possibility of substantive changes in national and transnational energy policies that draw their inspiration from practices being used on the ground. Eating fruit, vegetables, or even meat should not entail burning the amount of petroleum that is now being consumed. We can do much better than we are doing now.

At the end of the day, we must come to believe that all of these proposals are possible to carry out, even when others do not Our collective ecological circumstances are calling each one of us to a new level of faith, in our understanding of the Divine, of ourselves, of our neighbors, and of the biosphere. For this reason we must see that religion is not a disembodied set of beliefs or ritual but is in fact something that emerges from the lives of real people. Our faith traditions challenge us to see beyond the limitations of our individual imaginations, as well as our economic, political, and geographic circumstances. *Anyone* has the right to choose to be in league with a Creator God capable of making an entire universe. Everyone has the right to be an advocate for the sustainability of the biosphere. It is through such expressions of faith through action that we can become open to learning how to preserve and sustain the Creation we have been given the privilege of being born into. It is through faith that we can see the biosphere with new eyes and see ourselves as an interdependent part of its makeup. Such faith is not an exclusivist privilege, limited to any one religious tradition. Theology that claims an exclusive means to salvation must be rejected as unsustainable. The faith we are given begins with our specific ecological location, but it is not limited to any one place or any one people. Our faiths must ultimately be judged by the degree to which they call us to preserve the integrity of the biosphere. This is the earthly arbiter of our own tradition's efficacy. We cannot be afraid to imagine what we have not yet seen. The God of the Bible and the God of the Qur'an is the God of possibility. It is therefore up to us to remember who we are, where we live, and what we have been called to do. If Muslims and Christians choose to work together to preserve the Creation they claim to believe in, there is little they cannot accomplish, for together they make up over half of the world's population.

The Questions That Remain

There are a number of questions that persist in the wake of this study. It is the ultimate aim of this book to act as an invitation to others to take what has been presented here and push it further. While there are clearly more questions remaining than can be effectively posed at this time, there are a few that stand out and deserve special attention.

What is the real role of religion in international affairs?   The Johnston and Sampson book *Religion, the Missing Dimension of Statecraft* is clearly a watershed text. Nonetheless, the role that religion can play in future efforts at conflict resolution continues to be defined.[20] One difficulty is that there are still too few voices in the discipline of international relations that are willing to pick up where Johnston and Sampson left off. Despite the paucity of such voices, one thing appears clear: religion's primary role in the conduct of diplomacy is as a descriptive tool. Without considering the religious dimension of what is unfolding on the ground, those who form policy risk working from a position of blindness as to the modern motivations, the full history, and the authentic predispositions of the populations they hope to impact. What religions and the religious cultures produce cannot be separated from the people whom diplomats represent, anymore than the people can be separated from the land they inhabit.

If we accept that religion is first and foremost a descriptive means in the formation of diplomatic policy, then we must also entertain the strong possibility that it can be a predictive tool. Future diplomats must therefore consider both the dominant and marginalized religious beliefs and practices of a people when forming long-reaching policy. As was noted in chapter 1, such a practice holds the potential to save many a nation from the potential embarrassment that accompanies the assumption that all actors of importance are secular. Foreign policy often deals with people who are more foreign than many in the diplomatic community are willing to admit. The assumption that all leaders and their populations are secular and guided by Northern assumptions is no longer viable, nor was it ever. Even when considering what appears to be an increasingly secular North, a religious analysis remains critical for understanding the fuller picture. Ultimately, the role of religion in international relations theory and practice is an unfolding phenomenon. For this reason, future scholars and practitioners of diplomacy would be well advised to ally themselves with experts in the fields of religion, ethics, anthropology, and environmental science if they intend their work to retain relevance. The rules of engagement have changed forever, and it is time that the discipline of international relations reflected this fact.

Despite all these claims, many will question how effective the use of religious analysis could be in countries whose populations are thought to be largely "secular." Perhaps one should answer such a question with another question: is it possible for a country to truly be secular? In western Europe, it is not debatable that the core religious institutions that have long dominated the landscape are waning in their importance. Yet, despite this phenomenon, religion remains a highly potent prescriptive tool. The language of custom, of culture, and of ethics often remains rooted in what was the active religious tradition of a country and its people, long after its traditional religious institutions have fallen into disrepair as living communities. At the same time, secularism has not prevailed in every corner of any country. Smaller religious communities either as worshippers and/or activists, have retained vitality despite larger trends while new communities have emerged. Religion as it is practiced is constantly evolving. The fact that it is not necessarily evolving in predictable, traditional ways does not diminish its political importance and cultural potency. For this reason, diplomacy must also evolve to include a religious analysis among the central forms of describing the circumstances of a population, a nation-state, or bioregion. Thus, the role of religion can only be ignored at the risk of failing to grasp what is occurring on the ground in any location.

How do we move beyond religious exclusivism? Many religions' traditional claims to being the sole source of human salvation are among the most difficult barriers to achieving a deep level of cooperation between populations of different religious faiths. Christianity and Islam are the best demonstration of this truth. While many prefer to avoid this topic, the goal of long-term cooperation among individuals and communities of different faiths demands that this conflict be addressed directly.

Religious pluralism is both an opportunity and a challenge. When people of different faiths come together, there are always new opportunities for all sides to see the Divine in a different light. There is the possibility to engage in new forms of exegesis, with those of other traditions and other life experiences, with familiar texts and new ones. Regretfully, many such gatherings ignore difference and focus solely on commonalities. While such an approach can be productive in the short term, to expand existing relationships requires an examination of contradictory claims. Tolerance does not equal respect. Respect can only be earned as trust deepens.

What is the role of science versus the role of human self-restraint? While it is clear that many technologies have facilitated critical advances in such areas as agriculture, medicine, and education, it is also clear that science alone will not fortify, preserve, or create sustainable communities. The individual human often makes choices that undermine sustainability. Many in the human community believe that technology will save the biosphere. Still others believe that people should learn to turn off the lights when they leave a room, recycle glass, paper and metal, use public transportation, learn from those who are already live sustainably, and choose to use renewable resources. In many cases poor individual human choices still remain the greatest impediment to lasting sustainability. Science will not create a new more sustainable human. Individuals and collective communities must choose sustainability over greed, fear, and habit. How more sustainable human practices are to be brought about remains the subject of much debate.

What can help bring about the acceptance of the new currency of Ecological Realism? Many people remain convinced of the truth of traditional Realism, and much evidence supports this conclusion. In too many international agreements, weaker countries are forced to appease the wealthier ones or find their proposals made redundant. Traditionalism is also fortified by the false belief that the biosphere is in no real danger. For many in the "industrialized world" climate change and ozone holes are merely the concern of those who wish to diminish the efficacy of a free market global economy. Such "bad news" is labeled "false news," and allows many to rationalize their continued reliance on unsustainable practices.

Ecological Realism must become the new currency of an Earth economy that reflects real Earth value and the finite state of our collective natural resources. By using the Ecological Footprint as a measuring stick, any nation can begin to learn what it has and what it does not have to work with. This knowledge could become a cornerstone of public education and discourse with each individual household and community calculating its own Ecological Footprint. And just as the Euro was once an abstract currency, so too could Ecological Realism and the Ecological Footprint become prominent in future economic and political policy. For just as the Euro is regulated in Brussels and in Bonn, so too could Ecological Realism guide the decisions made by regulatory bodies. The first changes must come from individual communities, and then move to regional, national, and bioregional areas. Not so long ago, few believed that the currency of the EU could replace the centuries-old currencies of historical western Europe. Today it is an accepted fact. We need to have more faith that other even greater changes could take place in any corner of the world.

How can we lessen the tendency of the North to steal others' sustainability? This is a very difficult question to answer. In a free market economy, almost everything carries an artificial monetary price. Those who can meet the price can control any commodity within their grasp. Many industrialized nations have approached climate negotiations with the intention of purchasing the right to pollute from "less developed nations." The capital involved is not insignificant, and thus those on the economic margins are sorely tempted to trade their own clean air for a shot at rapid industrialization funded by the North. Somewhere, more people will eventually have to say "no" to the North's requests. As in Chiapas, more and more people in the South are coming to the realization that their economies are no longer under their own control. On the road to Ecological Realism this is an important realization to acquire. Southern nations who have changed their crops to suit the needs of their Northern clients are finding it more and more difficult to feed themselves. More people need to make this connection. When they do, changes that will have been previously inconceivable could become possible. All participants in a systemic diplomacy can come to the aid of countries who have become manufacturing and agricultural platforms for the North, and promote a return to more sustainable practices. Religious and secular NGOs can be one source of such advocacy. Dynamic national leaders can also play an important role. Northern control of Southern human and non-human resources is not necessarily permanent. Communities that represent religious Earth-centered traditions can provide a spiritual foundation that economic incentives lack. All of these actors working together could link ecological colonialism clearly and publicly to more traditional and reviled forms of colonialism. Such a multilevel strategy is crucial if current unsustainable practices are to be rejected by the affected populations.

Could the North learn to allow the South to share true ecological leadership, or perhaps even allow the South to lead the North to a more sustainable future? Such a proposition turns many time-honored traditions upside down. The economic North has long been accustomed to determining the rules of engagement and leading countries whose economies are unable to compete on Northern terms. Yet the wisdom required to enact long-term sustainability may not be found in the North. This is an argument articulated by Vandana Shiva and many other Southern thinkers.[21] At times, it appears that the arrogance of many Northern countries would preclude real ecological collaboration across the North–South split. Many in the North are comfortable with the fact that they have discarded centuries of ecological learning because such practices are seen as no longer meshing with the structure of a neoliberal economy. Still, there are many individuals in the North who part company with current Northern governmental and corporate leadership. These are the advocates of cultivating alternative energy sources, organic farming, local production for local consumption, and anti-sweat shop labor movements, among others. There is a lively amount of communication across the North–South split with the common goal of lessening our collective destruction of the human and nonhuman environment while promoting sustainability.

Ultimately, long-term progress can only be made if those in the economic North—particularly the United States, the EU, and Japan—concede that they are responsible for an inordinate amount of the biosphere's destruction. To appreciate the wisdom of existing sustainable communities will require an openness and a willingness to learn from others that the North has rarely demonstrated since before the expulsion of the Muslims from Spain in 1492. Today, more and more Northern historians will concede that were it not for the Muslims in Europe, many Northern advances in architecture, agriculture, medicine, philosophy, and mathematics would have been long delayed, or perhaps never occurred. What some historians are willing to admit, however, is not yet reflected in modern public policy. Learning to be the student when one has for so long played the role of master is very difficult. The most creative advocates of sustainability on both sides of the North–South split are called upon to determine how this transition could be made.

What does sustainability mean in an increasingly globalized world? According to the Christian theologian Julio de Santa Ana, "the socialized relationships which globalization favors are those which are often 'virtual' and less personal."[22] Thus, according to de Santa Ana, globalization "endangers community life by threatening people's relationships with their own cultures."[23] While such developments must clearly be resisted, globalization will not disappear anytime soon. Thus, the challenge of maintaining existing sustainable communities while building and expanding new ones must take place within the context of globalization itself. In this light, Julio de Santa Ana correctly acknowledges that one of the most important fights to change current consumption patterns will take place among the complacent and comfortable who "passively allow the market to assume the properties and dimensions of God in their imaginations and their behavior."[24]

In many modern cultures, wealth is a sign of the grace of God. Be it overt or implied, it is this type of theology that must be directly taken on if the negative effects of globalization are to be undermined. The Muslim theologian Farid Esack speaks eloquently to this point when he writes:

> This same text [Sura 5, known as Al-Ma'idah] links these notions of being God's favorites to their socio-economic implications and suggests that this sense of having an exclusive share of God's dominion leads to greater unwillingness to share wealth with others: "Have they perchance a share in Allah's dominion?" the Qur'an asks, and then asserts: "But [if they had] lo, they would not give to other people as much as [would fill] the groove of a date stone!"[25] (An-Nisa' 4:53)

Esack has hit on the same point articulated by de Santa Ana: those who have benefited the most from a globalized system will be the least likely to recognize the plight of those who have been left behind. An inherent theology of wealth implies that economic exclusion is the will of God.

Such an implied theology and eschatology must be discredited by the proponents of sustainability, with a more inclusive understanding of God and an entirely different definition of grace. Only truly inclusive and clearly articulated antiexclusionary theologies will help to fortify the foundations of sustainable communities. Sustainable diplomats must therefore be willing to turn perceptions on their head; naming greed and disgrace to be what many now call grace, while fighting all efforts to theologically or economically exclude those who live on the margins.

A New Call to Pilgrimage: Living Between the Two Trees

One of the most potent metaphors to describe current Muslim–Christian relations worldwide comes from the Muslim description of the Tree of Being as it was articulated by Ibn al-'Arabi. As Ibn al-'Arabi noted, we live between the fruits of the heavenly Tree of Being growing downward toward Earth and the fruits of its infernal upright counterpart. We live and make all of our most crucial decisions in this *saeculum*, or intersection between life as God has imagined it for us, and the hell that we daily invite into our lives and the lives of others. From which fruits will we eat from? Which will we offer our neighbor? Which will we export for an earthly profit? We are affected by what everyone eats. It changes our perception of everything—who we are, where we live, and the range of choices available to us. If we look upon the biosphere's current circumstances and truly believe that we are powerless to change anything of importance, what has become of our imaginations? Perhaps Julio de Santa Ana was correct when he wrote that we have made the market itself our God.[26] If we retain this perception, then *we are powerless* over the future of the biosphere.

The time that we live within the confines of the *saeculum* is inherently limited. Our time to change is even more limited. Now is the time to choose to see the connections to which we have been collectively blind—connections between people who live on either side of a human-made border, connections between people of different races and ethnicities, connections between people of different faiths, and connections between people of different levels of economic privilege and plight. The Sustainable diplomat must always be on the lookout for leveling influences, and collect a working transnational vocabulary from existing conversations. The Ecological Footprint is one excellent example of this. Recognition of new definitions of what it means to be a neighbor and to have neighbors can be realized through understanding the reality of bioregions, of genetic histories that transcend political borders, and through holding up the many common ethical principles different faith traditions hold to be true.

It would be delusional to claim that any of the above-mentioned changes will be easily realized. They fly in the face of how we currently understand most of our earthly realities. We are surrounded by circumstances, voices, beliefs, and systems that tell us that change is not within our grasp, or worse—that it is not even logical to contemplate. This is why we must celebrate those who are making changes despite any of these messages. Such changes may be small, from a Northern neighbor who sold her car and now rides the bus, to a Southern congregation who builds a church where the worshippers are joined under the same roof as the animals that sustain them. We must be diligent in always looking for new, creative models of sustainability. We must resist being disempowered by our governments, our employers, our lack of employment, our poverty, and the poverty of spirit found in so many who are economically rich.

Taking up a New Pilgrimage Together

Now is the time to reconsider one of the most time-honored rituals in both popular Christianity and Islam: the pilgrimage. This is because a pilgrimage is not merely a journey in search of a goal. A pilgrimage is every bit as much about what happens on the road, whom we encounter, whom we help, and who helps us. It is on this road that we encounter God, whether it is within the sea of Muslim humanity that annually makes the Hajj to Mecca, or the smallest Christian pilgrimage that is mounted in order to ask a virgin or a saint to intercede on someone's behalf—to talk to God together and raise our voices higher than we are able to do so alone.

The Christian and Muslim pilgrimages must merge. Somehow, together the Christians of Spain and the Muslims of Morocco must honor the salihs who honor the Earth. Somehow, together the people from both sides of the Strait must attend a *mousem*.[27] Somehow, together the Christians of Spain and the Muslims of Morocco must join in a *romeria*, and seek the assistance of a virgin or a saint who honors the Earth.[28] These are the rituals as they are practiced, as they have emerged from the lives of real people. These are rituals that have retained meaning for centuries despite the ebb and flow of official Christian and Muslim doctrine. Many of these pilgrimages end in a feast, where there is food for everyone who chooses to come. It is in the feast that we recognize our true *companions*, a word that in its Latin roots means "those with whom we share bread." We must do everything in our power to persuade *all* our neighbors to come, to sit, and to eat together. At the same time, we must do our utmost to visualize a Muslim salih and a Christian saint walking together. We must set aside conventional expectations, and try to imagine what they would say to one other.

The best diplomats can find a seat at the table for all participants, at a table that is inviting to everyone. Activist Christians and Muslims who work for Muslim–Christian reconciliation must now decide how to issue credible, mutual invitations. Practitioners of Sustainable Diplomacy will want to have a seat at this envisioned table: to listen, to learn, and to share everything they can bring to the meal. This is the goal of Sustainable Diplomacy—each side desiring that the other eat at its table as an honored guest, who in turn honors the host.

If we are truly creative, the talk of shared meals need not be relegated to the metaphorical level. In fact, individual exchanges such as those described have already taken place. They are models to be learned from and built upon. Invitations to eat at a common table must be made again and again until they are accepted as sincere. Sustainable communities eat together on a regular basis, because everyone must eat. The generosity embodied at the core of Christianity and Islam make such an exchange conceivable. The future of the biosphere, our common future, makes such a meal vital and necessary.

The Common Ground Beneath Our Feet

Recognizing our common biosphere is central to perceiving our common future. No one nation can secure clean air for itself without the cooperation of its neighbors. The quality and level of the cooperation required for such a project cannot be secured through military or economic intimidation. It will require popular movements that result in decisions by the majority of a population. Underground aquifers do not respect political borders. The reality of our mutual growing resource scarcities coupled with our growing pollution levels requires us to approach diplomacy in new and creative ways. In many cases, the land itself provides a means and a vocabulary that cuts across multiple divisions that separate human communities.

In the midst of our ecological crises, faith traditions that have emerged from the land can become revitalized. Supported by those who have kept faith traditions alive, practitioners of diplomacy have access to a living gift in the form of multiple insights into the lives of the people they hope to reconcile. There are new languages to learn and to use in conducting a diplomacy that is recognizable not only to leaders, but to the populations they represent. No diplomacy will be truly sustainable if it is not acknowledged as intelligible and valuable by general populations. Only general populations choosing to work together can make bioregions into political, economic, and social realities. Such a radical change regarding popular perceptions of bioregions cannot be simply declared from above. The ability to perceive bioregions as viable entities can only emerge from the lands and the faiths of the people who inhabit them.

Sustainable Diplomacy and Its Required Evolution

This book can only hint at the work ahead. Alone, it is simply a challenge to others to push further the concepts of Ecological Realism and Sustainable Diplomacy, determining for themselves what these terms mean in their own particular contexts. It is clear that the concepts of Ecological Realism and Sustainable Diplomacy must constantly evolve if they are to become and remain meaningful contributions to constructing a truly systemic diplomacy.

A sustainable future will require thinking further outside of the box than ever before. We must visualize a reality beyond the single nation-state. We must imagine a world where many more people take the concept of "Fair Earthshares" seriously. While we live in a postcolonial world in the traditional sense of the world "colonial," we must now confront the new colonialisms that often travel under the banner of our "inevitably and fully globalized" economies. Now we must confront ecological colonialism in a world of tremendous ecological illiteracy—its practitioners as well as those they have colonized. Current and future discourses must grapple with the successes and failures of those who seek independence from this cycle.

"There is a mountain between us," sings the Imazighn musician. It can be a mountain that separates, or a mountain that connects. All of us must open our eyes to the tools that Creation itself has provided for the task of promoting cooperation across borders. Those who seek change must enter into a deep willingness to learn from unconventional sources—from the land, and from people who do not share their faith, their race, their economic or political status, or even their hemisphere. The evolution of Sustainable Diplomacy requires turning the world as we know it on its head, and coming to realize that sweeping changes are not to be feared, but in fact embraced.

Appendix: Ecological Location Interview Questions[1]

Interview#_____ Date:_____ Location: _____
Name:
Age:
Gender:
Ethnicity:
Occupation:
Economic Class:
Level of Education:
City of Birth:
City of Residence:
Marital Status:
Additional Observations:

1. How long have you lived here? Where is your family from? Have the people in your family moved many times or stayed in the same place?
2. Why do you live in this village/city/region? Would you like to move? Where would you move if you were able? What would you miss if you moved?
3. What does it mean to you to live in a city/in a village/in a rural area? What do you like about the living in such a place?
4. Outside of your home, where is your favorite place to go/spend time? What is most beautiful to you in this village/city/region? Why?
5. Do you grow/raise the food that you eat? Where does it come from? Do you know how to raise vegetables and/or animals? Who taught you these skills? If you do not know how to do these things, who was the last person in your family who did such things? If you buy your food in the market, do you know where it comes from?
6. Outside of your family, what are the two or three most important groups of people in your life? What are the common interests which hold these groups of people together?
7. When you hear the word "land," or particularly the phrase "the land of Morocco/Spain,"[2] what images come to mind for you?
8. In your opinion, what is the most important animal that comes from your region? Plant? Tree? Why?
9. How do you feel connected to the land where you live? What is this connection made of? Do you own any land? Where is it?
10. Is your work connected to the land? How is this so/not so? How is your life affected if the rains do not come?

11. What do you see as your role in working for a good future for the next generation? Do you see their future as being connected to the health of the land? How is this so?

12. What do you think the land here was like when only the Imazighn/Iberians[3] lived here? What happened to the land when people came here from other places?

13. What do your religious and/or cultural traditions teach you about the origin of the land/Creation? Where did you learn about these traditions? Did you or will you teach these stories to your children? What do your religious and/or cultural traditions teach you about your connection to the land and other living creatures? Do you have special responsibilities regarding Creation because of your religious tradition? What are they?

14. In your religious tradition, is there a particular way of speaking about the beauty and holiness of the land? Is there a specific way in which you understand God's work in Creation? What do people say in your community? How do you speak of the blessings that come to you from the land?

15. What do you see as your connection, if any, to the people living in Spain/Morocco?[4] What are your biggest differences? Would you like to go to Spain/Morocco?[4] Why? What do you think are the most important things to do to improve Morocco and Spain's relationship? When you answer this last question, imagine that you are José Maria Aznar or Mohammed VI, and you had their authority.[5]

NOTES

Introduction: An Overview

1. Stanton Burnett, "Implications for the Foreign Policy Community," in Douglas Johnston and Cynthia Sampson, eds., *Religion, the Missing Dimension of Statecraft* (New York: Oxford University Press, 1994), 293.
2. The notion of analyzing relations between nation-states in a manner that privileges the role of religion is not new, but it remains peripheral to international relations theory.
3. Henry Munson, Jr., *Religion and Power in Morocco* (New Haven: Yale University Press, 1993), iv.
4. This anchoring theory of this inquiry is deeply indebted to the work of Daniel Spencer and his important contribution of the concept "Ecological Location" to the work of Christian Ethics. Spencer describes *ecological location* as "enlarging the term *social location* to include both where human beings are located within human society and within a broader biotic community. . ." Ecological Location, according to Spencer, also acknowledges that "how we are shaped to see and act in the world results from a complex interplay of physiological, social, cultural, *and environmental/ecological* factors. For ecological ethics (and, I would argue, the vast majority of social ethics), ecological location is the 'relevant whole' or context that must be taken into consideration in ethical reflection." While Spencer's application of Ecological Location focuses primarily on the individual or small group, this book seeks to expand and adapt the concept of Ecological Location to describe the circumstances of the overall population of a nation-state. See Daniel T. Spencer, *Gay and Gaia: Ethics, Ecology and the Erotic* (Cleveland, OH: The Pilgrim Press, 1996), 295, 296.
5. The phrase "on the ground," refers to people in the general population. It is a term that is employed throughout this text.
6. "Turning to the earth" is a concept proposed by Larry Rasmussen in his book, *Earth Community, Earth Ethics* (Maryknoll: Orbis, 1996), 110.
7. The interviews used in the course of this book were conducted from June through October of 2000 in six languages: Spanish, Derija (Moroccan Arabic), English, Tamazight, and Tachelhite (two of Morocco's principal indigenous languages) and Derija mixed with Hassania.
8. Included among those Christian Ethicists who have engaged the topic of religion and international relations are the following: John Bennett, *The Radical Imperative: From Theology to Social Ethics* (Philadelphia: Westminster Press, 1975); Alan F. Geyer, *Ideology in America: Challenges to Faith* (Louisville, KY: Westminster John Knox, 1997); Bryan J. Hehir, "Religion and International Affairs: Faith Can No Longer be Relegated to a Private Sphere in a World where State Sovereignty is Limited," *Nieman Reports* 47 (1993): 39–41; and Glen Stassen, *Just Peacemaking: Ten Practices for Abolishing War* (Cleveland, OH: Pilgrim Press, 1998).

9. Sustainable Diplomacy's normative guideposts are taken from Larry Rasmussen's requirements for sustainability. See Larry L. Rasmussen, *Earth Community, Earth Ethics* (Maryknoll: Orbis, 1996), 172–173. For a complete synopsis of Rasmussen's definition of sustainability, see David Wellman, *Sustainable Communities* (New York: World Council of Churches, 2001), 21–30.

Chapter 1   Interpreting Human Communities in Conflict

1. Song by Ali Ouzineb and Mohamed Qat, recorded in the original Tamazight in Ouaoumana, Morocco, July 12, 2000 by the author, translation from the original Tamazight by Sadik Rddad. Oral permission for recording this song was granted by Mr. Ouzineb and Mr. Qat on July 12, 2000 in Ouaoumana.
2. Ibid.
3. Many schools of international relations theory, including other forms of Realism, have emerged since in the inception of the Realist School. In the wake of Realism the Liberal and Constructivist schools have also come into their own.
4. Stanton Burnett, "Implications for the Foreign Policy Community," in Douglas Johnston and Cynthia Sampson, eds., *Religion, the Missing Dimension of Statecraft* (New York: Oxford University Press, 1994), 293.
5. Hans J. Morgenthau, *Politics Among Nations: The Struggle for Power and Peace* (New York: Alfred A. Knopf, 1978).
6. Ibid., 4.
7. Ibid.
8. Morgenthau writes, "The fact that a theory of politics, if there be such a theory, has never been heard before tends to create a presumption against, rather than in favor of, its soundness." Ibid.
9. Ibid., 5.
10. Ibid.
11. Ibid.
12. Ibid., 6
13. Ibid., 7.
14. Ibid., 8.
15. Ibid., 9.
16. Ibid.
17. Ibid., 10.
18. Ibid.
19. Ibid.
20. Ibid.
21. Ibid.
22. Ibid.
23. Ibid., 11.
24. Ibid.
25. Ibid.
26. Ibid.
27. Ibid.
28. Ibid., 12.
29. Ibid.

30. Ibid., 11.
31. Burnett, "Implications for the Foreign Policy Community," 293.
32. Ibid.
33. Morgenthau, *Politics Among Nations*, 364.
34. Douglas Johnston, "Introduction: Beyond Power Politics," in Johnston and Sampson, eds., *Religion*, 3.
35. Ibid., 5.
36. Edward Luttwak, "The Missing Dimension," in Johnston and Sampson, eds., *Religion*, 9–10.
37. Ibid.
38. Ibid., 12–13.
39. Barry Rubin, "Religion and International Affairs," in Johnston and Sampson, eds., *Religion*, 27.
40. Burnett, "Implications for the Foreign Policy Community," 294.
41. Daniel T. Spencer, *Gay and Gaia: Ethics, Ecology and the Environment* (Charlotte, OH: The Pilgrim Press, 1996), 295–296.
42. Ibid., 295.
43. Ibid., 295–296.
44. Ibid., 296.
45. Ibid.
46. Ibid., 299.
47. Ibid., 300.
48. See Mitchell Thomashow, *Ecological Identity: Becoming a Reflective Environmentalist* (London and Cambridge MA: MIT Press, 1995).
49. Spencer, *Gay and Gaia*, 314–315.
50. Ibid., 301–302.
51. Ibid., 313.
52. Mathis Wankernagel and William Rees, *Our Ecological Footprint: Reducing Human Impact on Earth* (Gabriola Island, BC, Canada: New Society Publishers, 1996), 13.
53. Ibid., 4.
54. Ibid., 9.
55. Ibid., 11.
56. Ibid., 14. At the time of this writing, the organization Redefining Progress had amended the per capita Fair Earthshare to be 2.1 hectares. It is this number that forms the basis of all subsequent calculations contained in this work.
57. Ibid., 32. This passage was excerpted by Wackernagle and Rees from the Brundtland Commission's text *Our Common Future* (New York: Oxford University Press, 1987), 27.
58. Ibid., 35
59. Ibid.
60. Ibid., 46.
61. Ibid.
62. Ibid., 44.
63. Ibid., 47.
64. Ibid., 68.
65. Ibid., 65.
66. The following equations are all taken directly from Wakernagel and Rees' text *Our Ecological Footprint*, 65–66.

67. Ibid., 65.
68. Ibid., 65–66.
69. Ibid., 66.
70. Ibid.
71. Ibid.
72. Ibid.
73. Morgenthau, *Politics Among Nations*, 11.
74. Larry Rasmussen, *Earth Community, Earth Ethics* (Maryknoll, NY: Orbis Books, 1996), 110.
75. It should be noted that many Creation stories have been historically used not to diminish, but to reinforce an anthropocentric view of the world. Work inspired by Eco-Realism will therefore require drawing on the work of those who have emphasized a rereading of scriptural texts with an eco-centric eye.
76. Wackernagel and Rees, *Our Ecological Footprint*, 46.
77. Larry Rasmussen defines sufficiency as "the commitment to meet the basic material needs of all life possible. This means sufficiency for both human and otherkind's populations. For humans it means careful organization of the economics of borrowing and sharing, with both floors and ceilings for consumption." See Rasmussen, *Earth Community Earth Ethics*, 172.
78. Ibid., 110.
79. Ibid., 343.
80. Ibid., 103.
81. Ibid., 260.
82. Ibid., 261, 284–285.
83. Ibid., 172–173. For a complete synopsis of Rasmussen's definition of sustainability, see David Wellman, *Sustainable Communities* (New York: World Council of Churches, 2001), 21–36.
84. Rasmussen, *Earth Community, Earth Ethics*, 42.
85. Wellman, *Sustainable Communities*, 23.
86. Ibid.

Chapter 2   The Foundations of the Eco-Historical Landscape of Moroccan–Spanish Relations

1. By "theological anthropologies" I am referring to the way a faith tradition places the human in regard to the Earth, nonhuman Creation, and to God.
2. Interview conducted in Seville, September 12, 2000 by the author and translated from the original Spanish by Deborah Avery.
3. Interview conducted in Fes, June 12, 2000 in English by the author.
4. Duncan Derry et al., *World Atlas of Geology and Mineral Deposits* (New York: John Wiley and Sons, 1980) 46, 50, 56.
5. Ibid., 50, 56.
6. The implications of this advantage could arguably be reflected in a number of ways regarding human technological development. For a detailed analysis of this phenomenon, see Jared Diamond, *Guns, Germs and Steel: The Fates of Human Societies* (New York: W. W. Norton, 1999). For as Diamond has argued, ". . . the striking differences between the long-term histories

of peoples of the different continents have been due not to innate differences in the peoples themselves but to differences in their environments." Ibid., 405.

7. *Spain: Geography*, Lexis-Nexis Academic Universe document, m=7a4ad94336cc3565eb8f, found at http://web..lexis-nexis.com/universe/document, December 10, 2001, 1.

8. Ibid., 2.

9. *Spain: An Environmental Overview*, Lexis-Nexis Academic Universe document, m=ff8e1e2caa4d39ae4360, found at http://web..lexis-nexis.com/universe/document, 12/10/01, 1. It should be noted that this statistic includes the rich biodiversity of the Canary Islands.

10. Ibid.

11. Also, see Arturo Ruiz and Manuel Molinos, *The Archaeology of the Iberians* (Cambridge: Cambridge University Press, 1998), 97.

12. *Spain: An Environmental Overview*, 1.

13. Ibid., 1.

14. L. Luca Cavalli-Sforza, Paolo Menozzi, and Alberto Piazza, *The History and Geography of Human Genes* (Princeton: Princeton University Press, 1994), 161.

15. Ruiz and Molinos, *The Archaeology of the Iberians*, 46.

16. Ibid., 50.

17. J. Donald Hughes, *Ecology in Ancient Civilizations* (Albuquerque, NM: University of New Mexico Press, 1975), 23.

18. Mark Ellingham and John Fischer, "The Historical Framework," *The Rough Guide to Spain* (London: Penguin, 1999), 833.

19. Hughes, *Ecology in Ancient Civilizations*, 23.

20. Ibid., 23–24.

21. Juan Lalaguna, *A Traveller's History of Spain, 3rd Edition* (New York: Interlink Books, 1996), 5.

22. Ibid.

23. Ibid.

24. Ruiz and Molinos, *The Archaeology of the Iberians*, 98.

25. Ibid.

26. Ibid.

27. Ibid., 100.

28. This area includes the contested Western Sahara, which Morocco currently controls. See "Country Profile: Morocco," *The Economist Intelligence Unit 2001* (London: The Economist, 2001), 17.

29. Though Morocco currently controls the Western Sahara, its ownership is highly disputed internationally.

30. This area includes the contested Western Sahara, which Morocco currently controls. See *Morocco: Geography*, Lexis-Nexis Academic Universe document, m=b00be376a700250dd92, found at http://web.lexis-nexis.com/universe/document, December 10, 2000, 2.

31. Dale Eickelman, *Moroccan Islam: Tradition and Society in a Pilgramage Center* (Austin: University of Texas Press, 1976), 15.

32. Abdelmalek Benabid, "Forest Degradation in Morocco," in Will D. Swearington and Abdedllatif Bencherifa, eds., *The North African Environment at Risk* (Boulder, CO: Westview Press, 1996), 175. To support this contention, Benabid cites the work of L. Emberger, "Une Classification biogeographique des climates," Rev. Trav. Lab. Bot. Zool. Fac. Sci.

Montpellier, Ser. Botan. 7 (1955): 3–43 and P. Daget, "Le bioclimate mediterraneen: caracteres generaux, modes de caracterisation," *Vegetatio* 34 (1977):
1–20 and Daget, P., "Le bioclimat mediterraneen: analyse des formes
climatiques par le systeme d'Emberger," *Vegetatio* 34 (1977): 87–103.

33. Eickelman, *Moroccan Islam* 15. Eikleman cites Daniel Noin as the source of
this information, see Daniel Noin, *La Population Rurale du Maroc, vol. 1–2*
(Paris: Presses Universitaires de France, 1970), 88–89.

34. Ibid.

35. *Morocco: Environmental Overview*, Lexis-Nexis Academic Universe document, m=6a1721fb0c268e983d27, found at http://web.lexis-nexis.com/
universe/document, 12/10/01, 2.

36. Ibid.

37. Ibid.

38. Ibid.

39. Cavalli-Sforza, *The History and Geography of Human Genes*, 161.

40. Ibid.

41. Cavalli-Sforza writes: "(The Iberomarusian's) type is sometimes referred to
as 'Mechta-Afalou.' Skeletal evidence indicates that these people were of
the Cro-Magnon type, that is, the same a. m. h. (anatomically modern
humans) found in Southwestern France and Spain." Ibid.

42. Ibid.

43. Ibid.

44. Ibid.

45. Michael Brett and Elizabeth Fentress, *The Berbers* (Cambridge, MA:
Blackwell Publishers, 1996), 11–12.

46. Ibid. There is other evidence of early exchanges between Iberian and
Imazighn peoples. Shell decorated pottery and bell beakers similar to
Iberian types were found in northern Morocco. See Brett and Fentress, *The
Berbers*, 15.

47. Abdallah Laroui, *A History of the Maghrib: An Interpretive Essay* (Princeton:
Princeton University Press, 1977), 19.

48. Ibid., 18.

49. Brett and Fentress, *The Berbers*, 35.

50. Ibid.

51. Ibid., see Mela Ponponius 1.8.45; G. Camps, "Funary Monuments with
Attached Chapels from the Northern Sahara," *African Archaeological Review*
4 (1986): 151–164.

52. According to Brett and Fentress, the Kings among the Imazighn were
accorded an even higher status in death: acting as intermediaries between
the living and the dead. See Brett and Fentress, *The Berbers*, 35.

53. Cavalli-Sforza, *The History and Geography of Human Genes*, 161.

54. Ibid.

55. I am using the term "Imazighn" as a general term to describe the numerous
indigenous tribes that inhabit Morocco. The term "Imazighn" is often the
preferred term among Morocco's indigenous peoples and the newly emerging indigenous scholars over the previously used Western term "Berber,"
which came from outside of Morocco, and had its origins in the Greek term
barbaroi, meaning "outsider" or "non-Greek." This same term gave birth to
the Engish term "barbarian" and thus took on numerous negative connota-
tions. To use the term "Imazighn" is thus to show respect for a people
who have chosen their own names, and reject a term that was imposed by

outsiders and colonizers. See Anthony G. Keen, in his review for the *Bryn Mawr Classical Review* October 14 (1994); Georges Pericles, *Barbarian Asia and the Greek Experience* (Baltimore: The Johns Hopkins University Press, 1994).

56. Brett and Fentress, *The Berbers*, 23.

57. Ibid., 23–24

58. Ibid., 24.

59. Cavalli-Sforza, *The History and Geography of Human Genes*, 161.

60. Ibid.

61. For more on this topic, see Bernal Martin, *Black Athena: The Afroasiatic Roots of Classical Civilization* (London: Free Association Books, 1987).

62. See "The First Europeans," *The Times of London*, March 20 (1997), Features Section.

63. Ibid. Alonso also posits that portions of Basque can be related to Etruscan, a language originally situated in Northern Italy.

64. Jose Ignacio Hualde, Joseba A. Lakarra, and R. L. Trask, *Towards a History of the Basque Language: Current Issues in Linguistic Theory, 131* (Philadelphia: John Benjamins Publishing Company, 1996), 88. While not specifying which Imazighn language he has extracted his words for comparison from, Hualde quotes the work of Anderson, which does offer some fascinating comparisons between a group of Basque and Imazighn words:

Basque		*Imazighn*	
izar	"star"	izeren	"star"
izn	"name"	ism/izn	"same"
aker	"male Goat"	iker/aker	"mutton"
anai(a)	"brother"	ana	"brother"
ama	"mother"	imma	"mother"
zamari	"horse"	zagmarz	"mare"
-a	(definitive article)	-a	(definitive article)

65. Antonio Tovar, "El vascuence y Africa," *Bolitin de la Real Sociedad Vascongada de los Amigos del Pais* 22, 3–4 (1966): 303–306, as cited in William A. Douglass, Carmelo Urza, Linda White, and Joseba Zulaika, eds., *Basque Cultural Studies* (Reno: University of Nevada, 1999), 31.

66. Abdallah Laroui, *A History of the Maghrib: An Interpretive Essay* (Princeton: Princeton University Press, 1977), 27. It was during this period, notes Laroui, that "Phoenicians, Greeks, Romans and Vandals entered the Maghrib, established settlements, and in some cases made their way far into the interior . . ." Ibid.

67. Cavalli-Sforza, *The History and Geography of Human Genes*, 287.

68. Sergio Ribichini, "Beliefs and Religious Life," in Sabatino Moscati, ed., *The Phoenecians* (New York: Abbeville Press, 1988), 104. According to Ribichini, Philo of Byblos "claims to have translated into Greek (the text) from the original Phoenician written by Sanchoniathon of Berytus, a priest who lived at the time of the Trojan War." Ibid.

69. Ibid., 105.

70. Ibid.

71. Ibid.

72. It is also interesting to note that what is called the "foundation myth" of Carthage names a woman as its founder. Known originally as Elissa, she was said to have earned the name "Dido" by the Libyans she first encountered

upon her arrival in Africa. See Serge Lancel, *Carthange: A History* (Cambridge, MA: Blackwell, 1995), 23.

73. Moscati, ed., *The Phoenicians*, 105.

74. Glen Markoe, *Phoenicians* (London: British Museum Press, 2000), 119–120.

75. It is fascinating to speculate upon the possible influence the Phoenicians had upon the religious beliefs of the Imazighn. Polytheism and the sacredness of particular places are all congruent with the prevailing understanding of ancient Imazighn beliefs and practices. Further, the cultural and political influence of the Phoenicians upon the Imazighn cannot be overestimated. Michael Brent and Elizabeth Fortress note that it was not long after the Phoenicians established the city of Carthage that the Imazighn were found to have adopted the title of "king," and its attendant political ramifications. The influence of Carthage was significant, as it eventually rivaled Greek and later Roman civilization in the western Mediterranean. Imazighn kingdoms emerged across the Maghreb, in many respects mimicking the political and cultural roles ascribed to Phoenician practices as projected by the rulers of Carthage. See Brett and Fortress, *The Berbers*, 24–25.

76. Many have hypothesized that the first Jews who arrived in the far west of the Maghreb did so by 361 B.C.E., over two centuries after the destruction of the first Temple in 586 B.C.E. See Daniel J. Schroeter, "Jewish Communities of Morocco: History and Identity," in Vivian B. Mann, ed., *Morocco: Jews and Art in a Muslim Land* (New York: Merril, 2000), 27.

77. Ibid.

78. Ibid., 28.

79. Ibid.

80. Ibid. Jewish communities persist to this day in all Maghrebi nations, though they have been significantly diminished by emigration. The presence of traditional Jewish trades such as silversmithing, goldsmithing, and moneylending have been greatly altered due to their practitioner's departure, often for Israel or the United States.

81. Hughes, *Ecology in Ancient Civilizations*, 77.

82. Ibid., 48.

83. Ibid., 49.

84. Ibid., 50.

85. J. Donald Hughes writes: "The Greeks were not always optimistic about changes wrought in nature, however. Heroditus . . . felt that many mighty works, like bridges and canals, were dangerous infringements on the natural order. In one of the best analyses in ancient times of human impact on the earth, Plato described the deforestation of Attica and the resultant soil erosion and drying of springs, so that 'what now remains compared with what then existed is like the skeleton of a sick man, all the fat and soft earth having wasted away, and only the bare framework of the land being left.' The conclusion that the earth under mankind's framework is undergoing degeneracy, not progress, was reached by many Greeks . . ." Ibid., 61.

86. Ibid., 71–72.

87. Ibid., 72.

88. For a detailed account of Roman mining techniques and their results, see Hughes, *Ecology in Ancient Civilizations*, 107–110.

89. Ibid., 102.

90. Ibid.
91. Ibid.
92. Ibid., 111.
93. Hughes writes: "The number of animals killed are phenomenal, mounting into the hundreds in a single day. Augustus had 3,500 animals killed in 3,500 *venationes*. At the dedication of the Colesseum under Titus, 9,000 were destroyed in 100 days, and Trajan's conquest over Dacia was celebrated by the slaughter of 11,000 wild animals." Ibid., 105.
94. Ibid., 103.
95. Ibid., 106.
96. Chew C. Sing, *World Ecological Degradation: Accumulation, Urbanization, and Deforestation 3000 B.C.—A.D. 2000* (New York: Alta Mira Press, 2001), 96.
97. Roman conservation efforts in some portions of their empire included the recycling of glass to minimize wood consumption, and even efforts to modify cooking techniques in order to use less wood in the preparation of meat. Ibid.
98. Hughes, *Ecology in Ancient Civilizations*, 87.
99. Ibid., 89.
100. Ibid.
101. Ibid., 128.
102. Ibid.
103. Herwig Wolfram, *The Roman Empire and Its Germanic Peoples* (Berkeley: University of California Press, 1997), 162.
104. Ibid.
105. Ibid.
106. J. M. Wallace-Hadrill, *The Barbarian West* (Oxford: Blackwell Publishers, 1996), 116.
107. Ibid.
108. Frank M. Clover *The Late Roman West and the Vandals* (Great Yarmouth, Norfolk: Variorum, 1993), 58.
109. Williston Walker and Richard A. Norris, David W. Lotz, and Robert T. Handy, *A History of the Christian Church, Fourth Edition* (New York: Charles Scribner's Sons, 1985), 148.
110. Ibid., 132.
111. Ibid., 133.
112. Clover, *The Late Roman West*, 666.
113. Ibid., 664.
114. Ibid., 167.
115. Ibid.
116. Brett and Fortress, *The Berbers*, 25.
117. Ibid., 76.
118. Walker et al., *A History of the Christian Church*, 148.
119. Andreas Schwarcz, "Cult and Religion Among the Tervingi and the Visigoths and Their Conversion to Christianity," in Peter Heather, ed., *The Visigoths: From the Migration Period to the Seventh Century—An Ethnographic Perspective* (Woodbridge: The Boydell Press, 1999), 447.
120. Ibid., 449.
121. Ibid., 450.
122. Wallace-Hadrill, *The Barbarian West*, 117.
123. Laroui, *A History of the Maghrib*, 76.

124. Walker et al., *A History of the Christian Church*, 231.
125. William C. Placher, *A History of the Christian Church: An Introduction* (Philadelphia: The Westminster Press, 1983), 92.
126. Williston Walker et al., *A History of the Christian Church*, 232.
127. See "The Historical Framework," in Mark Ellingham, ed., *The Rough Guide to Morocco* (London: Penguin Books, 1997), 490.
128. Lalaguna, *A Traveller's History of Spain*, 22. Henry Munson offered a fine synopsis of Sunni and Shi'i, when he wrote that "In the decades after the Prophet died in 632, there emerged three different views as to who should succeed him. The position that came to be associated with the [Shi'i] sects of Islam was that Ali [the Prophet's son-in-law and son of his uncle— Muhammad had no surviving sons of his own] and his patrilinial descendants should be the imams or 'leaders' of the Islamic community in part at least because they had inherited the Prophet's purity and infallibility (the name Shi'ite [or Shi'i] is derived from the term *shi 'at 'Ali*, 'the faction of Ali'). The position that prevailed among the people who came to be known as Sunni Muslims was that a *khalifa* (caliph) should be chosen by a council of the most influential men of the Islamic community. Although the Sunnis (named after the *Sunna*, or 'customary practice' of the Prophet) rejected the Shi'ite [Shi'i] view that only 'Ali and his descendants could rule the Islamic world, they did restrict the pool of potential caliphs to the men of the Prophet's tribe of Quraysh.'" Henry Munson, Jr., *Religion and Power in Morocco* (New Haven: Yale University Press, 1993), 35–36. Further, it should be noted that while the Umayyad Dynasty was based in Damascus, it launched its armies that reached the western Maghreb from Egypt.
129. Markus Hattstein and Peter Delius, *Islam: Art and Architecture* (Cologne: Konemann Verlagsgesellschaft, 2000), 208.
130. Laroui, *A History of the Maghrib*, 75.
131. Ibid.
132. Ibid., 92. For a more detailed account of the Khariji, see Hamid Dabashi, *Authority in Islam: From the Rise of Muhammad to the Establishment of the Umayyads* (New Brunswick: Transaction Publishers, 1989), 137–141.
133. Laroui points out, however, that "The conquest, which consisted essentially in the imposition of Arab sovereignty, meant neither Islamization nor Arabization. Arabization required many centuries and Islamization was the work of the Berbers themselves." Ibid., 87.
134. Ibid., 120–121. See also Frederick Mathewson Denny, *An Introduction to Islam, Second Edition* (New York: Macmillan Publishing Company, 1994), 164.
135. Denny, *An Introduction to Islam*, 91.
136. Tracing their roots to the Khajari, those who accepted the Barghwata originally came from the plains surrounding modern-day Casablanca. Their originator, a prophet by the name of Salih, revealed an alternative to the Qur'anic text in an Imazighn language. He was said to have been preached about by his grandson Yunus in the ninth-century C.E. Much of the appeal of the Barghwata and its followers was their promise to transform their followers (almost all Imazighn and marginalized) to people in a position of power vis-à-vis the Arabs. For more on this subject see "The Islamization of Morocco from the Arabs to the Almoravids," in Michael Brett, *Ibn Khaldun and the Medieval Maghrib* (Brookfield: Variorum, 1999), 66–67.

137. Laroui, *A History of the Maghrib*, 85.
138. Ibid.
139. Ibid., 86.
140. Juan Zozoya, "Material Culture in Medieval Spain," in Vivian Mann, ed., *Convivencia: Jews, Muslims, and Christians in Medieval Spain* (New York: George Braziler, Inc., 1992), 157.
141. The Idrissi Dynasty represents a phenomenon that would be repeated in Morocco: a hybrid group composed of both Arabs and Imazighn. Ultimately, many of the Imazighn would adopt Arabic as their primary language. This is a significant phenomenon in Morocco, a nation where one's ethnic identification is often defined by one's first language rather than one's bloodline.
142. Hattstein and Delius, *Islam*, 208.
143. Munson, Jr., *Religion and Power*, 40–41. The specific dates attached to each dynasty are drawn from William Spencer's *Historical Dictionary of Morocco* (London: The Scarecrow Press, 1980). I have placed term "Arab-Imazighn" in quotation marks beside the 'Alawi because they can arguably no longer be seen as a purely Arab dynasty. Morocco's current sovereign, Mohammed VI is of mixed blood, his mother rumored to be an Imazighn from Kenifra. This was the design of Mohammed VI's predecessor, Hassan II, who sought to unite the Arabs and Imazighn of Morocco and chose to reflect his goals in the makeup of his own family as well. While a patrilinial descent is favored among Muslims, the presence of Imazighn blood in the Alaw'i is accepted as common knowledge among Moroccans. The practice of intermarriage between Arabs and Imazighn is not new among Moroccan dynasties, as it was clearly the practice of the 'Alaw'i' predecessors, the Sa'dis, though its open acknowledgment is a more recent phenomenon. It should also be noted that the designation of the Sa'di and 'Alaw'i Dynasties as Sunni is based on their claim of descent from the Prophet and the Idrissis. It is however a matter of contention whether or not Idriss I was in fact Sunni or Shi'i.
144. Moses Maimonides, one of the greatest of all Jewish philosophers, theologians, scholars, and activists of all time, is one such example. Born in Cordoba in 1135, under the relatively benevolent reign of the Al-Andalous rulers, an independent branch of the Umayyads, he was forced to flee Iberia due to the intolerance of a later Muslim dynasty for Jews (the Almohads), who demanded conversion of the Jewish community to Islam or death. Maimonides, whose first language was Arabic, took an Arabic name and fled to the Maghreb, and later to the Levant. See M. Friedlander, "The Life of Moses Maimonides," in Moses Maimonides, *The Guide for the Perplexed* (New York: Dover Publications, 1956), xv–xxv.
145. Denny, *An Introduction to Islam*, 219.
146. Ibid., 220.
147. Ibid., 223.
148. Ibid., 225.
149. Ibid., 221.
150. Ibid., 222.
151. Laroui, *The History of the Maghreb*, 172.
152. Ibid., 192.
153. Norman Roth, *Jews, Visigoths and Muslims in Medieval Spain* (New York: Brill, 1994), 152.

154. Ibid.
155. Ibid., Roth documents such cases by citing the following: "Abu Zakariyya Ibn al-Awwam, *The Book of Agriculture* (ed. and trans. Jose Antonio Banqueri under the title *Libro de Agricultura* [Madrid, 1802], two volumes). On agriculture in general, see the article "Filaha," in Sanchez-Albornoz' *España Musilmana* I—especially the section on "Muslim West" (actually only al-Andalus); Lucie Bolens, "L'Agriculture hispano-Arabe au Moyen Age," with extensive bibliography; Glick, *Islamic and Christian Spain*, ch. 1; Arie, *Espana Musilmana*, ch. IV, Ibid., 294 ff.
156. Ibid.
157. Ibid. Roth cites Millas Vallicrosa's "La Traduccion Castellana del 'Trato de Agricultura' de Ibn Wafid," *Al-Andalus* 8 (1943): 281–332 and his *Traducciones Orientrales en los Manuscritos de la Biblioteca Catedral de Toledo*, Madrid (1942), 92 ff., as a critical source of material to varify these Muslim-inspired Christian advancements.
158. Bernard F. Reilly *The Medieval Spains* (Cambridge: Cambridge University Press, 1993), 91.
159. Reilly notes that Muslim dry farming techniques, as well as the water-wheel and the potchain, which raised the level of available water and thus the size of potentially arable land, were borrowed by the northern Spaniards from the Muslims, factors that served to increased an individual's ability to farm larger and larger tracts of land. Ibid., 91–92.
160. Ibid., 91.
161. Ibid., 92.
162. Ibid.
163. Ibid., 93. This act was initiated by the Council of Burgos in 1080.
164. Charles A. Truxillo, *By the Sword and the Cross: The Historical Evolution of the Catholic World Monarchy in Spain and the New World, 1492–1825* (Westport, CT: Greenwood Press, 2001), 29.
165. In fact, many of those who did convert later found the authenticity of their conversions to be questioned, a phenomenon that placed many former Jews and Muslims at the stake to be burned as heretics by the Inquisition.
166. L. P. Harvey, *Islamic Spain: 1250–1500* (Chicago: University of Chicago Press, 1992), 324.
167. Jaques Waardenburg is among those who ascribe to this analysis of the Reconquista. Waardenburg broadens the context, however, when he writes, "The question must be asked when and where the idea of such an absolute opposition [between Muslims and Christians] developed. When and where did it become an antithetical scheme which was imposed by both parties on all relationships between Muslims and Christians? Who made an ideology of it, generalizing what had been limited to specific historical and political situations? What has been the responsibility of theologians, whether Muslim or Christian? And when has the scheme of a general opposition between Muslims and Christians prevailed in European societies and public opinion? I am inclined to think of specific historical situations such as the period . . . of the Reconquista of Spain." See Jaques Wasardenburg, "Critical Issues in Muslim-Christian Relations: Theoretical, Practical, Dialogical, Scholarly," *Islam and Christian–Muslim Relations* 8, 1 (March 1997): 14.

168. Lalaguna, *A Traveller's History of Spain*, 259.
169. Richard Gillespie, *Spain and the Mediterranean: Developing a European Policy Towards the South* (New York: St. Martin's Press, 2000), 66.
170. Ibid.
171. For an excellent and detailed timeline of these and other Spanish historical events, see Lalaguna, *A Traveller's History of Spain*, 256–264.
172. Adrian Shubert, *A Social History of Modern Spain* (New York: Routledge, 1996), 100–101.
173. Ibid.
174. Lalaguna, *A Traveller's History of Spain*, 189.
175. Shubert, *A Social History of Modern Spain*, 234.
176. According to Mary Nash, "National Catholicism was the vehicle for cultural and political indoctrination from childhood to adulthood. The church-controlled school system guaranteed the dissemination of Catholic doctrine. According to law, teaching was based on Catholic doctrine, while of course, religion was an obligatory subject. Not only Catholic dogma, but religious imagery, iconography and rituals shaped the education, culture and [cosmological vision] of all Spaniards throughout their life. . . . José Pemartín, one of the leading ideologues of National Catholicism, openly spoke of the effectiveness in harnessing popular Church cults to political interests: 'Think of our . . . Cult, the most magnificent and marvelous on earth; our processions and pilgrimages, fiestas, saints' celebrations, May stations and splendid processions at Corpus, Holy Week; think of fascist Spain, the State has to live . . . much more intensely with all the people.' Franco himself made maximum use of religious rituals and imagery, such as the recurrence [of] the famous arm of Saint Teresa of Avila to enforce his image of authority and divine legitimacy." See Mary Nash, "Towards a New Moral Order: National Catholicism, Culture and Gender," in José Alvarez Junco and Adrian Shubert, eds., *Spanish History since 1808* (New York: Oxford University Press, 2000), 291–292.
177. Seen as an alternative to American liberalism or Soviet Communism by its apologists, *corporatism* was viewed by Franco as a "third way," a system that is "a political world view derived from Aristotle and St. Thomas Aquinas: that government is good and natural; that it need not, however, be checked and balanced; that the well-ordered political system is integrated, disciplined, hierarchical; that all groups and individuals are secure and fixed in their station in life. . . . In Spain and Portugal the main historic 'corporations' or corporative units of society, traced back to feudal or even earlier times, where the family is seen as the building block of society, the local community or town, the parish or neighborhood association, the Catholic Church, the military orders, the guilds of sheepherders, silversmiths, and other trades, and the university or religious orders. Typically, supreme authority in a corporatist regime is centered in a governing body based not on geographic representation (one person, one vote) but on functional representation from the groups indicated above. Because they derive their ideas so strongly from medieval Catholic political philosophy, corporative regimes of this kind have been particularly prevalent in the Catholic countries of Europe (Spain, Portugal, Austia, Italy, Belgium, even France) and Latin America." See Howard J. Wiarda and Margaret

MacLeish Mott, *Catholic Roots and Democratic Flowers: Political Systems in Spain and Portugal* (Westport, CT: Praeger, 2001), 42.

178. Ibid., 36.
179. Shubert, *A Social History of Modern Spain*, 209.
180. Ibid.
181. Ibid., 209–210.
182. Laroui, *A History of the Maghrib*, 105.
183. Ibid., 142. By the fourteenth century, Morocco had become a great exporter of many agricultural items, as well as human beings. According to Laouri, "The foreign merchants chiefly frequented the Mediterranean ports; on the Atlantic they went no further than Salé. Morocco exported slaves, leather, carpets, cereals, sugar, and coral. It imported, as did Ifriquiya, wines, cloth, and metals. These goods were carried largely by European ships . . ." Ibid., 217.
184. Ibid., 234.
185. Originally, the Sufist movement in Morocco was aimed at democratizing and deepening the Islamic faith. Yet as Sufism failed to construct itself as a centralized force, the movement developed increasingly autonomous groups. See Laoroui, *A History of the Maghrib*, 245. It must be noted, however, that positing a concrete syncretism between Sufism and pre-Islamic practices remains a contested hypothesis. See Vincent J. Cornell, *Realm of the Saint: Power and Authority in Moroccan Sufism* (Austin: University of Texas Press, 1998). It should also be noted that the use of the term "caliph" (meaning "deputy," "successor," or "replacement") originally referred to the four men who led the Islamic Umma after the death of the Prophet. The term "Sultan," on the other hand, is defined by Henry Munson as being "authority" or "government." Despite the decided difference in the meaning of these two terms, many sultans have claimed for themselves the title of caliph or khalifa, in an effort to link themselves to the family of the Prophet and/or claim religious importance beyond their original scope of influence. See Henry Munson, *Religion and Power in Morocco*, 37–39.
186. Vincent Cornell holds that the claim of the existence of a Maraboutic crisis is in and of itself specious. He argues that the simple presence of Sufi activism does not in and of itself comprise a challenge to any sultan's rule. Rather, those painted with the brush of rebellion were in fact doing their utmost to preserve Islam in Morocco. Cornell states, "Once again, the role of the saint appears to be of a different order than that of the sage or spiritual master. As both imam and salih, the Moroccan saint of the early modern period was more than just a teacher or a mystic. Instead, he symbolized all aspects of the Muhammadan paradigm. As a spiritual master, he imitated the prophet Muhammad as an interpreter of religion and a model of piety; as a salih, he imitated Muhammad as a social critic and a friend of the poor; as an imam, he imitated Muhammad as a leader of the people. In the Jazulite model of sainthood as summarized in the doctrine of at-tariqa al-Muhammadiyya, Moroccan shaykhs found a means to manifest both the social and the religious dimension of the prophetic archetype in a single persona. This is what made them so influential in political affairs. For the shaykhs of the Jazuliyya, the involvement of saints and Sufis in the political conflicts of the time was no 'maraboutic crisis' motivated by a desire for personal gain. Instead, these exemplars of the prophetic

Inheritance stepped into a preexisting leadership vacuum and did their best to preserve the integrity of Muslim society according to the dictates of their calling." See Cornell, *Realm of the Saint*, 233.

187. For a position that affirms the possible existence of the Maraboutic Crisis, see Eickelman, *Moroccan Islam*, 15–29.

188. Laroui, *A History of the Maghrib*, 245–246.

189. Tamazight and Tachelhite are today two of the most prominent Imazighn languages spoken in Morocco.

190. Laroui, *A History of the Maghrib*, 253. In some cases, such a system benefited the general population, but in many instances it did not, as in the case of the sugar trade between England and Chefchaouen. Ibid., 257.

191. This account contradicts many accounts offered by colonialist historians, who hold that the products of Morocco, apart from wheat, were not in great demand in the North. See Cornell, "Socioeconomic Dimensions of the Reconquista and Jihad in Morocco," *Journal of Middle East Studies* 22 (1990): 379–418.

192. Ibid., 383.

193. Ibid.

194. Laroui, *A History of the Maghrib*, 271. For more details, see Cornell, "Socioeconomic Dimensions of Reconquista and Jihad in Morocco: Portuguese Dukkala and the Sa'Did Sus, 1450–1557," *International Journal of Middle East Studies* 22, 4 (November 1990): 396.

195. Laroui, *A History of the Maghrib*, 271.

196. Ibid., 275.

197. Ibid., 276.

198. Ibid., 281.

199. Ibid., 317.

200. It was also during this period that Moroccan currency was replaced by Spanish and French coinage. See C. R. Pennel, *Morocco Since 1830: A History* (New York: New York University Press, 2000), 80.

201. Laroui, *A History of the Maghrib*, 318.

202. Pennell, *Morocco Since 1830*, 166.

203. Ibid., 167.

204. Ibid., 168. In many cases, the "Moroccan recruits" turned out to be Imazighn. This strategy went hand-in-hand with the French's stated policy of cultivating a strong relationship with the indigenous peoples of Morocco against the Arab leadership. At prior points in Moroccan history, as it has been noted, this attempt at cleaving the indigenous from the Arab found the French contemplating the strategic advantage of attempting to convert the Imazighn to Christianity, thus cementing their separation from the rest of the Moroccan population and making them more dependent upon the Europeans. This historical attempt at European-engineered ethnic separation is critical in understanding modern tensions between the Arab and Imazighn populations of Morocco. See Pennell, *Morocco Since 1830*, 212–213.

205. Ibid., 187.

206. Ibid., 169.

207. Laouri, *A History of the Maghrib*, 328.

208. Ibid., 341. Laouri states: "Colonial violence merely severed the few remaining ties between the historical domain (states, cities, Islamic

justice and ritual) and the infrahistorical (*zawiyas*, rural communities, customs, folklore and private life)." Ibid., 345.

209. Pennell, *Morocco Since 1830*, 171. While the protectorate took primacy over Moroccan institutions, one of its first acts was to exempt Europeans from Islamic law, and create a French administered court system. Ibid., 173.

210. Ibid., 171.

211. Laouri, *A History of the Maghrib*, 330. Writes Laouri: "In Morocco, from 1927 on, the *malk* (privately owned) lands were expropriated (62,000 hectares in the Tadla in two years) and ceded to Europeans. In the Maghrib as a whole 3,800,000 hectares—roughly a third of the land under actual cultivation—had by 1930 passed into the hands of Europeans, who at no time made up more than a seventh of the total population." Ibid., 331.

212. Ibid., 367. The military resistance took place in the Rif, where the Riffis challenged Spanish control, and in more than one instance humiliated their invaders. What came to be known as the "Riffian Rebellion" was the most tangible form of Moroccan resistance to colonialism in the first half of the twentieth century. However, as the Riffis pushed toward French-controlled territory, they accomplished the feat of uniting the French and the Spanish against them. See Pennell, *Morocco Since 1830*, 188–195.

213. Munson, Jr., *Religion and Power in Morocco* 93. The Riffian wars stood as a humiliation against the Spanish, one they would not soon forget. Costing the Spanish well over 10,000 troops simply in their retreat from Chaouen in 1924, the indigenous Riffis of the mountains showed the Spanish that they were not simply mercenaries for European use. Nor were they simple fodder for the guns of Franco. In turn, these same fiercely independent people would later provide a formidable armed opposition to King Hassan II, cementing the fact that the occupants of the Rif resented political domination in all forms, be they from Europe or their own land-mass. See Pennell, *Morocco Since 1830*, 188–192.

214. Pennell, *Morocco Since 1830*, 185. Those who were to attempt to use religion to undermine the power of the colonialists were of the Salafi school of Islam. Writes Pennell, "In 1921, Salafi activists founded several schools in Fez, . . . Others were soon set up in Rabat, the new capital, Salé, and Casablanca. In 1924 the first school was founded in Tetuán. Others followed in Marrakech and Kénitra . . . [Many of the leaders in this edu-cational movement] would later become important nationalist leaders, but for the moment they concentrated on education. They called their establishments 'Free Schools' not because the pupils paid no fees, but because they were free of French control." Ibid., 186.

215. Ibid., 183. Yet efforts to separate the Imazighn from the Arabs by the French were often met with resistance. There are recorded instances of Arabs praying openly in the mosques not to be separated from their "Berber brothers." Ibid., 213.

216. Ibid., 200. These were difficult times for Morocco not only due to European efforts. Pennell writes, "In early 1929, locusts ate the leaves of two thirds of the fruit trees in the Sous, and by the end of the autumn drought had killed a third of the cattle and a fifth of the sheep. In the spring of 1930 locusts returned for a second season." Ibid., 211.

217. Ibid., 205.

218. Ibid., 231. The actual title "king" would not be used officially until Moroccan independence in 1956.

219. Ibid., 244.
220. Ibid., 253.
221. Ibid., 254.
222. Ibid., 255. As Pennell notes, the already dire circumstances in Morocco were heightened by the collapse of the harvest in the Spanish zone in 1940, a phenomenon followed by four successive poor harvests. Ibid., 258.
223. Ibid., 262.
224. Ibid., 284.
225. This is not to say that the Spanish abandoned all their holdings in Morocco. They retained the northern fortified cities of Ceuta (or Sebta to the Moroccans) and Melilla, along with a large swath of the Western Sahara.
226. According to C. R. Pennell, naming Sidi Mohammed "king" freed the new Moroccan leader from the limitations of his predecessors: "Sidi Mohammed was no longer a sultan, but a king, a modern sovereign, who would be able to repudiate the past, and in particular the Treaty of Fez, that assigned him the title of Sultan and so confined everything Moroccan to the traditional, the oriental and the antique. While the title did not become official until independence . . . this was the first stage in resolving the duality between the Moroccan and the modern roles of the state." Pennell, *Morocco Since 1830*, 231.
227. Ibid., 295.
228. Ibid., 299.
229. Ibid., 303–304.
230. Ibid., 313.
231. Hassan's success at perpetuating the Alawi Dynasty lies in part in his ability to connect himself to the power of Islam, the celebrations remembering the life of the Prophet, and his deft use of television and radio to do so. As "Commander of the Faithful" Hassan named himself as the principal Imam for Morocco's population, and as a descendent of the Prophet. He did not lose any opportunity to remind the Moroccan population of his role as the continuation of a sacred lineage. Hassan, of course, is merely extending the role established for him by his long-ruling family, the Alawi. The Alawi Dynasty itself owes a great deal of its success to its initial ability to take for itself the legacy left by the previous sharif dynasty, the Sa'di. What the Sa'di founded in terms of its sherifi caliphate status, the Alawis built upon to the betterment of its own perceived basis of legitimacy. Presented not as an aberration, but as a continuum, the Alawis positioned themselves to benefit heavily from already established traditions and expectations, while simultaneously casting dispersions on their predecessors. See, M. E., Combs-Schilling, *Sacred Performances: Islam, Sexuality and Sacrifice* (New York: Columbia University Press, 1989), 172–173, 178–180.
232. Pennell, *Morocco Since 1830*, 324. Pennell writes: "By 1960 Morocco became a net importer of cereals and has remained so ever since." Ibid., 325.
233. Ibid., 358. Morocco's applications to join the EC (which later became the EU) were rejected. However, it is interesting to note that there has been talk of the possibility of one day building a tunnel under the Strait of Gibraltar between Morocco and Spain. While the plan has been put on hold, it still captures the imagination of many, particularly on the Moroccan side of the Strait. Tunnelbuilder.com notes, "Of various

pro-posals put forward over the years, the one first to result from inter-governmental discussions between Morocco and Spain envisage(d) a 50 km tunnel, with 28 km under the seabed with ground cover of up to 100 m. The route is from Punta Paloma to Tangiers, relieving the port of Algeciras, which is grossly overloaded at times as workers move between North Africa and the European Union. The 1992 cost estimate was Pta 2,000 million." For more tunnel related information, see info@tunnel-builder.com.

234. This view was adhered to despite numerous human rights abuses under Hassan's rule by the Moroccan Army, as well as the secret police.

235. Like the Idrissis before him, Hassan II has perpetuated the old Moroccan tradition of marriage between an Arab male and an Imazighn female, or so many in Morocco believe. Photos of the wives of Hassan II were not permitted, though many average Moroccans note the strong Imazighn features in King Mohammed VI's face, a factor that only lends credence to this theory. Whether or not such actual or perceived genetic ties between Arabs and Imazighn can be capitalized upon by the new king has yet to be played out. It can be said that many of the young king's first acts were seen as favorable to Morocco's Imazighn population. Further, Imazighn languages such as Tamazight, Tachelhite and Riffi can now be found on Moroccan national television, and it is now considered by many to be only a matter of time before they are taught in the majority of Moroccan universities. In a break with tradition, Mohammed VI announced his intentions to marry a young woman from Fes with a degree in computer sciences. Her photos were officially released to the Moroccan media, and the Moroccan people are now looking forward to what would appear to be a more public royal family. Mohammed VI is still young and enjoys significant popularity among the majority of his constituents. Whether the bloom will fade in this relationship between sovereign and subjects ultimately has yet to be seen. Much will be expected of this new king, particularly among his generational peers, many of whom cannot find work despite their holding university degrees.

Chapter 3   The Conflict Over Land: The First Human, Land Use and the Two Cities

1. Interview conducted in Seville by the author October 9, 2000 and translated from the original Spanish by Deborah Avery.

2. Interview conducted by Hamid Nouamani in Khenifra, recorded by the author, June 6, 2000 and translated from the original Tamazight by Sadik Rddad.

3. While it is a temptation to point out, for example, the differences in the Creation accounts of Genesis I and II and the way one story has been used in an abusive manner (the Genesis II Eve being created from Adam's rib), a canonical reading of the biblical text that presents these stories as a unified whole is essential if they are to be compared to the Qur'an. This is because in all but a few approaches to interpreting the Qur'an, the text received by Muhammad is considered by the vast majority of Muslims to be a direct and unimpeachable account of God's words, and thus it cannot be subjected to source, rhetorical, or historical criticism.

4. M. E. Combs-Schilling, *Sacred Performances: Islam, Sexuality and Sacrifice* (New York: Columbia University Press, 1989), 223. Combs-Schilling reminds her readers repeatedly that Hassan II as the King of Morocco tied his own life to that of the Prophet's by the ritual use of key Muslim traditions in which he placed himself at the center: the Id al-Kabir and the Prophet's birthday are two such examples.

5. All biblical citations are taken from *The New Revised Standard Version, The New Oxford Annotated Bible* (New York: Oxford University Press, 1994).

6. All Qur'anic citations are taken from Ali Ahmed, *Al-Qur'an: A Contemporary Translation* (Princeton: Princeton University Press, 1988).

7. **Genesis 3:19**, "But by the sweat of your face you shall eat bread until you return to the ground, for out of it you were taken; you are dust, and to dust you shall return."

    Al-A'raf 7:24–25, "Go," said God, "one the enemy of the other [Iblis (or 'Satan') and the humans], and live on the earth for a time ordained, and fend for yourselves. You will live there, and there you will die," [God] said, "and be raised from there (on the Day of Doom)."

8. For more on the Qur'anic interpretation of Adam as God's reflection, see Ibn Al'Arabi, *The Bezels of Wisdom* (Mahwah, NJ: The Paulist Press, 1980), 50.

9. One of the Qur'an's most well-known interpreters writes, "The Reality wanted to see the essences of His Most Beautiful Names or, to put it another way, to see His own Essence, in an all-inclusive object encompassing the whole [divine] Command, which qualified by existence, would reveal to Him His own mystery." Ibn al-'Arabi as translated by R.W.J. Austin, "The Wisdom of Divinity in the Word of Adam," *The Bezels of Wisdom*, 50.

10. Ibn al-'Arabi writes, "Concerning the knowledge of the Reality we say that it is eternal, whereas of [humanity's] knowledge we say that it is contingent." Ibid., 53.

11. The statistics cited are drawn from an amalgamation of three principal sources: the database of the World Resource Institute, entitled *World Resources 2000–2001: People and Ecosystems, the Fraying Web of Life* (Washington DC: United Nations Development Programme, United Nations Environment Programme, World Bank, World Resources Institute, 2001); The World Wildlife Fund and their calculations of the Ecological Footprint by M. Wackernagel, A. Callejas, and D. Deumling for Redefining Progress, found at www.panda.org/livingplanet/lpr00/downloads/lpr_2000_eco_crop_graze.pdf + ecological + footprint&hl = en; and statistics provided by members of the Redefining Progress team and their website at www.rprogress.org/programs/sustainability/ef/, 2/20,02. All final calculations for the ecological footprints presented in this text were compiled and calculated by Nova Gutierrez and the author. Readers should know that the majority of statistics presented in 2001 source texts represent 1996 statistics, the last year in which substantive statistics cutting across the board have been definitively compiled as of this writing.

12. Country: Spain
    Land area: 49,944,000 ha
    Population: 39,627,600
    Methods of calculations and conversions were provided by Mathis Wackernagel and William Rees, in their book *Our Ecological*

Footprint: Reducing Human Impact on the Earth (Gabriola Island, BC: New Society Publishers, 1996). Those calculations and conversions not found in the Wackernagel and Rees text were provided by the organizations The World Wildlife Fund and Redefining Progress, 2/21/02.

I. ENERGY LAND (SPAIN)
    Land required to sequester the CO_2 emitted in the atmosphere
A. Using Spain's CO_2 emissions, calculate amount of land that would be required to absorb those emissions.

CALCULATIONS
232,484,500 tons CO_2 emitted (Spain)
Ratio used for comparison (constant based on world average): 1 ha
1.8 tons of carbon emitted each year
232,484,500 tons $CO_2 \div 1.8$ tons $CO_2 = 129,158,056$ ha required to absorb Spain's CO_2 emissions

Per capita CO_2 emissions
129,158,056 ÷ 39,627,600 people = **3.2 ha/cap** (Ecological Footprint for Spain's Energy)
B. Land available to Spain for CO_2 absorption.
    Total forests in Spain: 8,388,000 ha
    8,388,000 ha ÷ 39,627,600 people = **.21 ha/cap** (Ecological Footprint for land available for Spain's CO_2 absorption)

13. II. CONSUMED LAND (SPAIN)
    Degraded land—built up land (settlements and roads),
    CALCULATIONS
    497 m²/capita
    CONVERSIONS
    m² × .0002471054 = acres
    acres ÷ 2.471054073 = ha
    1 ha = 10,000 sq. ms
    497 sq. meters/cap ÷ 10,000 = **.0497 ha/cap** (Ecological Footprint for Spain's consumed land)

14. III. CURRENTLY USED LAND (SPAIN)
    Crop Land, Managed Forests, Pastures, and Gardens
A. Using data on production/consumption data and avg. yield, calculate currently used land.
    a. Crop Land
       Arable and Permanent Crop Land   19,164,000 ha
       1. Cereals
          Yield   3,387.3 kg/ha
          Global yield   2,641 kg/ha
          Total Production   22,196,300 tons
          Cereal Donations   33,600 tons
          Total Cereal   22,162,700 tons
       CONVERSIONS
       1 ton = 907.1848 kg
       22,162,700 × 907.1848 = 20,106,000,000 kg
       20,106,000,000 kg ÷ 2,641 kg/ha = 7,612,898.4 ha

7,612,898.4 ÷ 39,627,600 = **.19 ha/cap** (Ecological Footprint for Spain's cereal's consumption)

2. Roots & Tubers
Yield 22,245.7 kg/ha
Global yield 13,385 kg/ha
Total Production 3,214,500 tons
3,214,500 tons × 907.1848 = 2,916,100,000 kg
2,916,100,000 ÷ 13,385 = 217,866.68 ha
217,866.68 ÷ 39,627,600 = **.005 ha/cap** (Ecological Footprint for Spain's roots & tubers consumption)

3. Pulses
Yield 719.8 kg/ha
Global yield 772 kg/ha
Total Production 382,800 tons
Imports 636,300 tons
Exports 13,500 tons
Total Pulses 1,005,600 tons
1,005,600 × 907.1848 = 912,270,000 kg
912,270,000 ÷ 772 kg/ha = 1,181,690.5 ha
1,181,690.5 ÷ 39,627,600 = **.03 ha/cap** (Ecological Footprint for Spain's pulses consumption)

4. Vegetables & Fruit
Global yield 12,120 kg/ha
Total production 24,338,000 tons
24,338,000 × 907.1848 = 22,079,000,000 kg
22,079,000,000 kg ÷ 12,120 = 1,821,704.9 ha
1,821,704.9 ÷ 39,627,600 = **.05 ha/cap** (Ecological Footprint for Spain's vegetables & fruits consumption)

5. Tobacco
Global yield 1,602 kg/ha
76,490 tons × 907.1848 = 69,390,565 kg
69,390,565 ÷ 1,602 = 43,314.96 ha
43,314.96 ÷ 39,627,600 = **.001 ha/cap** (Ecological Footprint for Spain's tobacco consumption)

6. Sugar
Global yield 3,229 kg/ha
1,467,660 × 907.1848 = 1,331,400,000 kg
1,331,400,000 ÷ 3,229 = 412,337.83 ha
412,337.83 ÷ 39,627,600 = **.01 ha/cap** (Ecological Footprint for Spain's sugar consumption)

b. Managed Forests
Total roundwood production 15,631,000 m³
Roundwood exports 660,101 m³
Roundwood imports 2,370,000 m³

Wood fuel prod. 3,198,000 m³
Industrial roundwood prod. 12,433,000 m³
Sawnwood prod. 3,080,000 m³

Wood-based panel prod.   2,970,000 m³
Paper and paperboard prod.   4,196,000 metric tons

CALCULATIONS
Roundwood Consumption
15,631,000 – 660,101 = 14,970,899 + 2,370,000 = 17,340,899 m³
(Tot. roundwood prod. minus exports plus imports = total round-
wood consumption)
Total saw wood & wood-based panels = 6,050,000 m³
Paper and paperboard production = 4,196,000 metric tons

CONVERSIONS
acres × .4046856 = ha
acres × 4046.856 = sq. m
1 hectare = 10,000 square meters
17,340,899 + 6,050,000 = 23,390,899 m³ (roundwood prod. + other
totals)

Paper and Paperboard
The production of each metric ton of paper requires 1.8 m³ of wood
4,196,000 metric tons × 1.8 m³ = 7,552,800 m³ (total wood required
for paper)

Total forest Used for Consumption
23,390,899 + 7,552,800 = 30,943,699 m³
2.3 m³/ha/yr. (avg. forest yield)
30,943,699 ÷ 2.3 = **13,453,782 ha**
13,453,782 ÷ 39,627,600 = **.34 ha/cap** (Ecological Footprint for
Spain's forest consumption)

  c. Pastures
    1. Meat Consumption Per Capita
      111.4 kg/cap × 39,627,600 = 4,414,500,000 kg
      4,414,500,000 ÷ 810 kg/ha = 5,450,018.1 ha
      5,450,018.1 ÷ 39,627,600 = **.14 ha/cap** (Ecological Footprint for
      Spain's pastures meat consumption)
    2. Milk Consumption
      7,434,480 tons
      7,434,480 × 907,1848 = 6,744,400,000 kg
      6,744,400,000 ÷ 336 kg/ha = 20,072,760 ha
      20,072,760 ÷ 39,627,600 = **.5 ha/cap** (Ecological Footprint for
      Spain's milk consumption)
      Total: (Cropland .286) + (Managed Forests .34) + (Pastureland
      .64) = **1.266 ha/cap** (Ecological Footprint of Spain's Currently
      Used Land)

  15. According to Redefining Progress, the CCD defines desert as hyper-arid,
      semi-arid, and semi-humid lands. Given that both the people of Morocco
      and Spain have in many instances successfully farmed semi-humid lands,
      and there are as yet no hard statistics for hyper-arid land, our desert space
      was therefore calculated by adding arid and semi-arid lands. Desert statis-
      tics provided by telephone and e-mail by Dr. Robin White, Senior
      Researcher, World Resources Institute, February 23, 2002.

16. IV. LAND OF LIMITED AVAILABILITY (SPAIN)
    Untouched Forests (Productive Natural Ecosystems), Non-Productive Areas (Deserts, Icecaps)
    CALCULATIONS
    a. Deserts
       Arid   4,301 sq. km
       Semi-Arid   193,598 sq. km
       Total   197,899 sq. km
       CONVERSIONS
       $(1,000 \text{ m})^2 = (1 \text{ km})^2$
       $1,000,000 \text{ m}^2 = 1 \text{ km}^2$
       1 hectare = 10,000 sq. m ÷ 1,000,000 sq. m = .01 sq. km.
       197,899 sq. km. ÷ .01 km. = 19,789,900 ha
       19,789,900 ha ÷ 39,627,600 = **.499 ha/cap** (Ecological footprint for deserts)
    b. Untouched Forests (Productive Natural Ecosystems—Not Managed)
       CALCULATIONS
       (Forests) Total roundwood production: 15,631,000 m^3
       Total sawnwood, wood-based panels = 6,050,000 m^3
       4,196,000 metric tons × 1.8 m^3 = 7,552,800 m^3 (total wood required for paper)
       15,631,000 + 6,050,000 + 7,552,800 = **29,233,800 m^3** (total forest used for production)
       2.3 m^3/ha/yr. (avg. forest yield)
       29,233,800 ÷ 2,3 = 12,710,348 ha
       12,710,348 ÷ 39,627,600 = **.32 ha/cap** (Ecological Footprint for Forest used for Wood Production)
       Closed forest   11,731,000 ha
       Plantation   1,925,000 ha
       Shrubs and trees   12,611,000 ha
       Total forest   26,267,000 ha
       Forest used for wood   12,710,348 ha
       Forest not used for wood   13,556,652 ha
       13,556,652 ÷ 39,627,600 = **.34 ha/cap** (Ecological Footprint for Untouched Forests)
    c. Polar and Alpine
       998,880 ha ÷ 39,627,600 people = **.03 ha/cap** (Ecological Footprint of Spain's Polar and Alpine Land)
       Total: (Deserts .499) + (Untouched Forests .32) + (Polar and Alpine Land .03) = **0.849 ha/cap**
       (Ecological Footprint for Spain's Land of Limited Availability)

17. THE ECOLOGICAL FOOTPRINT OF SPAIN
    General Summary

    Spain's Ecological Footprint
    4.5 ha/cap × 39,627,600 = 178,320,000 ha
    Spain's actual area   49,944,000 ha
    128,376,000 ha
    128,376,000/39,627,600 = 3.24 ha/cap (hectares per capita "over the limit" based on country size)

2.1 ha is the amount of land allotted to every human on earth if we were to all live equally. The amounts by which one has exceeded one's limit are calculated based on the size of one's individual country.
SPAIN'S ACTUAL ECOLOGICAL FOOTPRINT = 4.5 − 2.1 World share = 2.4 OVER FAIR EARTHSHARE

18. I. ENERGY LAND (MOROCCO)
    Land required to sequester the CO_2 emitted in the atmosphere

    C. Using Morocco's CO_2 emissions, calculate amount of land that would be required to absorb those emissions.
    CALCULATIONS
    27,879,400 tons CO_2 emitted (Morocco)
    Ratio used for comparison (constant based on world average): 1 ha
    1.8 tons of carbon emitted each year
    27,879,400 tons CO_2 ÷ 1.8 tons CO_2 = 15,488,556 ha required to absorb Morocco's CO_2 emissions
    Per capita CO_2 emissions
    15,488,556 ha ÷ 37,376,700 people = **.41 ha/cap** (Ecological Footprint for Morocco's Energy Land consumption

19. Land available to Morocco for CO_2 absorption.
    Total forests in Morocco : 4,079,000 ha
    4,079,000 ha ÷ 37,376,700 people = **.11 ha/cap** (Ecological Footprint for land available for Morocco's CO_2 absorption)

20. II. CONSUMED LAND (MOROCCO)
    Degraded Land—Built Up Land (Settlements and Roads)

    CALCULATIONS
    79 m²/capita (built-up land per person)
    CONVERSIONS
    m² × .0002471054 = acres
    acres ÷ 2.471054073 = ha
    1 ha = 10,000 square meters
    79 sq m/cap ÷ 10,000 = **.0079 ha/cap** (Ecological Footprint for Morocco's Consumed Land)

21. III. CURRENTLY USED LAND
    Crop Land, Managed Forests, Pastures, and Gardens

    B. Using data on production/consumption data and avg. yield, calculate currently used land.

    a. Crop land
       Arable and Permanent Crop Land 9,595,000 ha

    1. Cereals
Yield	1,123.8 kg/ha
Global yield	2,641 kg/ha
Total Production	66,333,000 tons
Cereal Receipts	3,500 tons
Total Cereal	66,336,500 tons

CONVERSIONS
1 ton = 907.1848 kg
66,336,500 × 907.1848 = 60,179,000,000 kg
60,179,000,000 kg ÷ 2,641 kg/ha = 22,786,000 ha
22,786,000 ÷ 37,376,700 = **.61 ha/cap** (Ecological Footprint for Morocco's cereals consumption)

2. Roots & Tubers

Yield	19,035.9 kg/ha
Global yield	13,385 kg/ha
Total Production	1,129,800 tons

1,129,800 tons × 907.1848 = 1,024,900,000 kg
1,024,900,000 ÷ 13,385 = 76,573.581 ha
76,573.581 ÷ 37,376,700 = **.002 ha/cap** (Ecological Footprint for Morocco's roots and tubers consumption)

3. Pulses

Yield	584.9 kg/ha
Global yield	772 kg/ha
Total Production	225,900 tons
Imports	24,500 tons
Exports	10,700 tons
Total Pulses	239,700 tons

239,900 × 907.1848 = 217,450,000 kg
217,450,000 ÷ 772 kg/ha = 28,167 ha
28,167 ÷ 37,376,700 = **.008 ha/cap** (Ecological Footprint for Moroccco's pulses consumption)

4. Vegetables & Fruit

Global yield	12,120 kg/ha
Total production	5,093,831 tons

5,093,831 × 907.1848 = 4,621,000,000 kg
4,621,000,000 kg ÷ 12,120 = 381,274.43 ha
381,274.43 ÷ 37,376,700 = **.01 ha/cap** (Ecological Footprint for Morocco's vegetable & fruit consumption)

5. Tobacco

Global yield	1,602 kg/ha

19,636 tons × 907.1848 = 17,813,481 kg
17,813,481 ÷ 1602 = 11,119.526 ha
11,119.526 ÷ 37,376,700 = **.0002 ha/cap** (Ecological Footprint for Morocco's tobacco consumption)

6. Sugar

Global yield	3,229 kg/ha

991,970 × 907.1848 = 899,900,000 kg
899,900,000 ÷ 3,229 = 278,693.13 ha
278,693.13 ÷ 37,376,700 = **.007 ha/cap** (Ecological Footprint for Morocco's sugar consumption)

b. Managed Forests

Total roundwood prod.	1,746,000 m³
Roundwood exports	100 m³
Roundwood imports	225,000 m³

Wood fuel prod.	770,000 m^3
Industrial roundwood prod.	976,000 m^3
Sawnwood prod.	83,000 m^3
Wood-based panel prod.	34,900 m^3
Paper and paperboard prod.	110,000 metric tons

CALCULATIONS

Roundwood consumption

1,746,000 − 100 = 1,745,900 + 225,000 = 1,970,900 m^3

(Tot. roundwood prod. minus exports plus imports = total roundwood consumption)

Total sawnwood & wood-based panels = 117,900 m^3

Paper and paperboard production = 110,000 metric tons

CONVERSIONS

acres × .4046856 = ha

acres × 4046.856 = sq. m

1 hectare = 10,000 sq. m

1,970,900 + 117,900 = 2,088,800 m^3 (roundwood prod. + other totals)

Paper and Paperboard

The production of each metric ton of paper requires 1.8 m^3 of wood

110,000 metric tons × 1.8 m^3 = 198,000 m^3 (total wood required for paper)

Total forest Used for Consumption

2,088,800 + 198,000 = 2,286,800 m^3

2.3 m^3/ha/yr. (avg. forest yield)

2,286,800 ÷ 2.3 = 994,260.87 ha

994,260.87 ÷ 37,376,700 = **.03 ha/cap** (Ecological Footprint for Morocco's consumption of managed forests)

  c. Pastures

    i. Meat Consumption Per Capita

      19.4 kg/cap × 37,376,700 = 725,110,000 kg

      725,110,000 ÷ 810 kg/ha = 895,195.04 ha

      895,195.04 ÷ 37,376,700 = **.02 ha/cap** (Ecological Footprint for Morocco's meat consumption)

    ii. Milk Consumption

      965,887 tons

      965,887 × 907,1848 = 876,240,000 kg

      876,240,000 ÷ 336 kg/ha = 2,607,900 ha

      2,607,900 ÷ 37,376,700 = **.07 ha/cap** (Ecological Footprint for Morocco's milk consumption)

      Total: (Cropland .6372) + (Managed Forests.03) + (Pastures .09) = **0.7572 ha/cap** (EF of Currently Used Land)

22. This measurement does not include the contested Western Sahara, an area which is currently only officially recognized as part of Morocco by the Moroccan government.

23. IV. LAND OF LIMITED AVAILABILITY:
    Untouched Forests (Productive Natural Ecosystems), Non-Productive Areas (Deserts, Icecaps)

CALCULATIONS

a. Deserts

Arid   112,446 sq. km.
Semi-Arid   191,615 sq. km.
Total   304,061 sq. km

CONVERSIONS

$(1,000 \text{ m})^2 = (1 \text{ km})^2$
$1,000,000 \text{ m}^2 = 1 \text{ km}^2$
1 ha = 10,000 sq. m ÷ 1,000,000 sq. m = .01 sq. km.
304,061 sq. km. ÷ .01 km. = 30,406,100 ha
30,406,100 hectares ÷ 37,376,700 = **.81 ha/cap** (Ecological Footprint for deserts)

b. Untouched Forests (Productive Natural Ecosystems—Not Managed)

CALCULATIONS

(Forests) Total roundwood production: 1,476,000 m^3
Total sawnwood, wood-based panels = 117,900 m^3
110,000 metric tons × 1.8 m^3 = 198,000 m^3 (total wood required for paper)
1,476,000 + 117,900 + 198,000 = **1,791,900 m^3** (total forest used for production)
2.3 m^3/ha/yr. (avg. forest yield)
1,791,900 ÷ 2,3 = 779,086.96 ha
779,086.96 ÷ 37,376,700 = **.02 ha/cap** (Ecological Footprint for forest used for wood production)
Closed forest   1,455,000 ha
Open forest   1,091,000 ha
Plantation 490,000 ha
Shrubs and trees   1,265,000 ha
Total forest   4,301,000 ha
Forest used for wood   −1,791,900 ha
Forest not used for wood   2,509,100 ha
2,509,100 ÷ 37,376,700 = **.067 ha/cap** (Ecological Footprint for untouched forests)

c. Polar and Alpine

446,300 ha ÷ 37,376,700 = **.01 ha/cap**
Total: (Deserts .81) + (Untouched Forests .067) + (Polar and Alpine .01) = **.887 ha/cap** (Ecological Footprint for Morocco's Land of Limited Availability)

24. MOROCCO'S ECOLOGICAL FOOTPRINT **1.2 ha/cap**

3 37,376,700 = 44,852,040 ha
Morocco's actual area −44,630,000 ha
= 222,040 ha
222,040/37,376,700 = **.006 ha/cap** (hectares per capita "over the limit" based on country size)
MOROCCO'S ACTUAL FOOTPRINT = 1.2–2.1 Fair Earthshare = 2.9 BELOW FAIR EARTHSHARE
2.1 ha is the amount of land allotted to every human on earth if we were to all live equally. The amounts by which one has exceeded one's limit are calculated based on the size of one's individual country.

25. Interview conducted and recorded in Seville by the author, September 9, 2000 and translated from the original Spanish by Deborah Avery.
26. Interview conducted in Fes by Muhammad Oukili, recorded by the author, June 16, 2000, and translated from the original Derija by Kamal Mzoughi.
27. Richard Gillespie, *Spain and the Medditerranean: Developing a European Policy Towards the South* (New York: St. Martin's Press, 2000), 66.
28. From an interview conducted and recorded by the author, September 14, 2000 in Seville, and translated from the original Spanish by Deborah Avery.
29. Interview by Muhammad Oukili, recorded by the author, June 16, 2000 in Fes, and translated from the original Derija by Kamal Hassani and the author.
30. This is a fact that was repeatedly borne out through the interviews conducted for this study, as well as by the population figures of rural vs. urban in each respective country. See World Resources Institute's Earthtrends Database, at earthtrends.wri.org, "Population, Health and Human Well-being—Urban and Rural Areas: Urban population as a percent of total population," pdf format, Morocco, 6, Spain, 8, 6/17/03.
31. See *World Resources 2000–2001: People and Ecosystems Data Base* (Washington DC: United Nations Development Programme, United Nations Environment Programme, World Bank, and the World Resource Institute, 2001) file # 21235.wki and #23620.wki.
32. See World Resources Institute's Earthtrends Database, at earthtrends. wri.org, "Population, Health and Human Well-being," Morocco, 6, Spain, 8.
33. The anthropologist Dale F. Eikelman provides an important explanation as to the difference between "God's will" in this circumstance and the long-popular false French colonial inclination to equate a belief in "God's will" with a belief in fate. States Eikelman: "God's will legitimates the present—and ephemeral—distribution of social honor as the God-given state of affairs. The ranking of individuals in relation to one another is never taken for granted but is constantly empirically tested. Provisionality is the very essence of the cosmos. Consequently, attention is focused upon assessing exact differentials of wealth, success, power, and social honor among *particular* men as a prelude to effective, specific action, not upon speculation over the general order of the world. God reveals his will through what happens in the world, and [people] of reason constantly modify their own courses of action to accommodate this will." See Dale F. Eikelman, *Moroccan Islam: Tradition and Society in a Pilgrimage Center* (Austin: University of Texas Press, 1976), 126.
34. This theme will be examined in depth in chapter 5, which examines the phenomenon of immigration between Morocco, Spain, and greater Europe.

Chapter 4   The Conflict Over Natural Resources: The Tree of Life and the Tree of Being, the Consumption of Natural Resources, and the Fish Wars

1. Interview conducted and recorded by the author, September 27, 2000 in Barcelona, and translated from the original Spanish by Peter Huijing.
2. Interview conducted by Brahim Ouajjani and recorded by the author, July 9, 2000 in Ait Abdellah Tahjala Ait Oufka Tafrout, and translated from the original Tachelhite by Salam Benzidi.

3. Muhyi-D-Din Ibn Al-Arabi, *Shajarat al-Kawn*, trans. A. Jeffery (Lahore, Pakistan: Aziz Publishers, 1980), 34. All subsequent commentaries by Ibn al-'Arabi on the Tree of Being will be drawn from this text. It should be noted that while this text is widely attributed to Ibn al-'Arabi, such an attribution does not represent a consensus among scholars.

4. Ibid., 36–37.

5. Evan Eisenburg, in his book *The Ecology of Eden*, writes, " 'So [God] drove out the [human]: and [God] placed at the east of the garden of Eden cherubim, and a flaming sword which turned every way, to keep the way of the tree of life.' The Tree of Life is the inner core of the world-pole: the heart of the heart of the world. [The hu]man must be prevented from reaching— and ruining—the source of life." See Evan Eisenburg, *The Ecology of Eden* (New York: Alfred A. Knopf, 1998), 97. Eisenburg goes on to write: "What exactly is the fiery sword? Is it our awe of the wilderness? Our fear of its rigors and dangers? Our discomfiture in the face of its unearthly beauty? Whatever it is, it is the best friend we have. For only by keeping our distance from wilderness—some wilderness at least—can we keep from fouling the wellspring of our own life." Ibid.

6. Ibn al-'Arabi writes: "If, therefore, you look at how the shoots of the Tree of [Being] differ, and at the kinds of fruit it has, you will recognize that the root thereof springs up from the seed of *kun* [being], separating out from it. So when Adam entered the School of Instruction and was taught all the names, he looked at the similitude of *kun*, looking at what [God] Who brings into being had proposed should be brought into being, and saw that what was taught by the *K* of *kun* was the *K* of treasure (*kanziyya*)." Ibn al-'Arabi, *Shajarat al-Kawn*, 30.

7. Writes Richard: "This is the utopia of those prevented from eating and drinking and from leading a secure life, those without money. Now in Revelation this is the ultimate life, guaranteed by God beyond death and oppression . . . In the new Jerusalem there now appears the tree of life that God offered in [God's] life-giving design for humankind in Genesis 2:9. When humanity chooses the project of death, it loses access to the tree of life (Gen 3:34). At this point the tree of life is now seen to be producing fruit twelve times a year. God's project of life for humankind is achieved in the church. The leaves from these trees serve as medicine to heal the nations that were sick as a result of Babylon's idolatry." Pablo Richard, *Apocalypse: A People's Commentary on the Book of Revelation* (Maryknoll, NY: Orbis Books, 1995), 164–165.

8. Writes Al-Arabi: "The shoots of the tree differ, and its fruits are of different kinds, that there may appear to the sinner the mystery of [God's] forgiveness, of [God's] mercy towards [those] who do good, [God's] kindness to the one who is obedient, [God's] favor to the believer and [God's] vengeance on the unbeliever." Ibn al-'Arabi, *Shajarat al-Kawn*, 34.

9. Ibid., 33. Ibn al-'Arabi writes: "The Divine Throne was set as a treasure house for this tree and as a store house for its arms, a house from which is to be sought the succour of all the goodness it contains. In it are the attendants and the servants of this tree."

10. Richard, *Apocalypse*, 145.

11. Ibn al-'Arabi, *Shajarat al-Kawn*, 48–49.

12. Ibid., 30.

13. Ibid., 29.

14. Richard, *Apocalypse*, 144.

15. Ibid., 164.

16. Ibn al-'Arabi, *Shajarat al-Kawn*, 33.

17. Ibid., 37 ff.

18. Redefining Progress and World Wildlife Fund Ecological Footprint data, primarily based on 1996 statistics, located on the world wide web at www.panda.org/livingplanet/lper000/downloads/lpr_2000eco_crop_graze. pdf, April 25, 2002. Additional information was generously supplied by Dr. Chad Monfreda at Redefining Progress.

19. Ibid.

20. Ibid., this statistic includes seafood imported by Spain.

21. Ibid.

22. Ibid.

23. Ibid.

24. Ibid.

25. Ibid.

26. Ibid.

27. Ibid.

28. Ibid.

29. Interview conducted in Fes by Muhammad Oukili, recorded by the author, June 13, 2000, and translated from the original Derija by Kamal Hassani and the author.

30. Interview conducted and recorded in Seville by the author, September 14, 2000, and translated from the original Spanish by Deborah Avery.

31. Richard Gillespie, *Spain and the Mediterranean: Developing a European Policy Toward the South* (New York: St. Martin's Press, 2000), 45.

32. Ibid.

33. Ibid.

34. Ibid.

35. Ibid., 46.

36. Ibid., 47.

37. Ibid.

38. Ibid.

39. Ibid.

40. Ibid., the *Cortes* is the national bicameral Parliament of the Spanish government.

41. Ibid., 47–48.

42. Ibid., 48.

43. Ibid., 50.

44. Ibid.

45. Ibid.

46. "Morocco: European Commission Demand the Resumption of Agricultural Trade," *Al-Hayat*, February 25, 2002. Found via lexis-nexis. com/universe/document?_m=2fdfa2dec17ce4a8e0f266 . . . May 21, 2002.

47. "Roundup: Diplomatic Crisis Worsens Between Spain and Morocco," Deutsche Presse-Agentur, February 1, 2002. Found via lexis-nexis.com/ universe/document?_m=7bec75affo5cf8dde13a1e . . . May 21, 2002.

48. This statistic is based on Redefining Progress' estimate of the Biological Capacity of Spanish fishing grounds, a footprint of .04 ha/cap, in stark contrast to the consumption footprint of .56 ha/cap.
49. "Mauritania: Mauritania, EU Renew a 5-Year Fishing Accord," *National Trade Data Market Reports*, September 25, 2001. Found on Lexis-Nexis at universe/document?_m=b71ba3d79ac5042f2249edc54d41638f&_docnum =17&wchp=dGLSlV-lSlzV&_md5=8ade6b775290a5dd1cd1cobe39bf5946, May 24, 2002.
50. "Country Profile: Spain," *The Economist Intelligence Unit 2001* (London: *The Economist*, 2001), 37.
51. This effort, among others, has led Spain to find herself entangled in other "fish wars" not dissimilar to the one she has long maintained with Morocco. This fact is well documented in an article published by *The Financial Times*, reprinted from *The Santiago Times* on July 19, 2001, which states:

> The National Fishing Society (Sonapesca) released a study denouncing the grave consequences of granting European fishing vessels free access to ports in countries such as Canada, Argentina, Morocco and Mauritania. The report intensifies local businessmens' fears associated with Chile's potential agreement with the European Union (EU) giving European fishermen free access to local ports, which is a key part of the possible Free Trade Agreement between the EU and Chile. Argentina exported US $ 1 billion worth of fish products before it signed an agreement with Spain granting free access to its ports in the early 1990s. The agreement allowed Spanish fishing vessels to extract 590,000 tons of hake per year. This dropped to 312,000 tons some years later due to over-exploitation. These catastrophic consequences led Argentina to withdraw free access to its ports in 1999. Canada experienced a similar problem in its cod industry. Six years after signing an agreement with Spain giving fisherman a fixed quota for cod catches, Canadian cod production collapsed. The crisis in the industry led to the so-called cod war, in which the Canadian fishing industry criticized the agreement because Spanish vessels are largely subsidized. Similar agreements with EU countries have devastated local fishing industries in some African countries. Chile's fishing industry exports US $ 1.8 billion annually. Authorities have had to establish restrictions on extraction to help protect mackerel and anchovy populations from over-exploitation.

> Drawn from the Lexis-Nexis database at http:universe/document?_m= 54d2b5653bfe372ff2c819db3e3662fe&wchp=dGLSlV-lSlzV&_ md5=aa2e2bdf24b50afe5a362411c69e097e, May 22, 2002.

52. "Country Profile: Morocco," 34.
53. Ibid., 29.

Chapter 5   The Conflict Over People: The Story of Abraham and Ibrahim and the Strangers, the Consumption of Illegal Human Labor and the Conflict Over Immigration

1. Interview conducted and recorded by the author, September 11, 2000, in Seville and translated from the original Spanish by Deborah Avery.

2. Interview conducted by Muhammad Ouikili and recorded by the author, July 14, 2000, in Fes and translated from the original Derija by Kamal Hassani and the author.

3. Muhyi-D-Din Ibn al-'Arabi, *The Bezels of Wisdom* (Mahwah, NJ: Paulist Press, 1980), 95.

4. Ibid., 91. See the introductory notes by R. W. J. Austin.

5. Imam Imâduddin Abdul-Fida Isama'il Ibn Kathîr Ad-Damishqi, *Stories of the Prophets* (Riyadh: Darussalam, 1999), 157. This is a text that is among a number of widely read *tafsir* (interpretation) in the Muslim world, which explicate key portions of the Qur'an.

6. Augustine writes: "Thus both kinds of [humans] and both kinds of households alike make use of the things essential for this mortal life; but each has its own very different end in making use of them. So also the earthly city, whose life is not based on faith, aims at an earthly peace, and it limits the harmonious agreement of citizens concerning the giving and obeying of orders to the establishment of a kind of compromise between human wills about the things relevant to mortal life. In contrast, the Heavenly City—or rather that part of it which is on pilgrimage in this condition of mortality, and which lives on the basis of faith—must need make use of this peace also, until this mortal state, for which this kind of peace is essential, passes away." See Augustine of Hippo, *The City of God* (London: Penguin Books, 1984), 877.

7. While it is in the Christian tradition that the sacrificial son is Isaac, for the Muslims, it is Ishmael (or Ismai'il). This reflects an entirely different take on Abraham's relationship with his offspring. While in the Jewish and Christian traditions, Hagar and her child are banished and left to fend for themselves, the Qur'an teaches that Abraham maintained two separate households, and supported Sarah, Hagar, and all of their children.

8. This Qur'anic position is underscored by the passage in Al-Anbiya' (21:91–94), where it is written: "(Remember) her who preserved her chastity, into whom We breathed a life from Us, and made her and her son a token for (humankind). Verily this your order is one order, and I am your Lord; so worship Me. But they split up the order among themselves; (yet) all of them have come back to me. So he who does the right and is a believer, will not have his labor denied, for we are cognizant of it." For further thoughts on this theme, see William C. Chittick, *Imaginal Words: Ibn al-'Arabi and the Problem of Religious Diversity* (Albany: State University of New York Press, 1994).

9. This point is drawn from a telephone conversation with Dr. Chad Montefreda of Redefining Progress on 6/6/02.

10. J. A. Rodriquez and S. F. Fuentes, "Spain: Riot Town Immigrants Extend Deadline on Strike," *El Pais*, February 26, 2000, as cited *Migrant News* 7, 3 (March 2000) at http://migration.ucdavis.edu/mn/archive_mn/mar_2000-11mn.html, June 19, 2003.

11. Ibid.

12. Marlise Simmons, "Resenting African Workers, Spaniards Attack," *New York Times* (February 12, 2000): A3.

13. It should be noted that there were two other murders that shortly preceded the one that sparked the riots in El Ejido, both allegedly committed by Moroccans. Ibid.

14. "E. U.: Migration Policy?" as cited in *Migration News* 7, 8, August 2000 at http:/migration.ucdavis.edu/mn/archive_mn/aug_2000-08mn.html, June 19, 2003.

15. Giles Tremlant, "Spain Sets Migrant Radar Trap," *The Times*, August 18, 2000, as reported in the *Migration News* 7, 9 (September 2000) at http://migration.ucdavis.edu/mn/archive_mn/sep_2000-12mn.html, June 19, 2003.

16. Ibid.

17. Elizabeth Nash, "Spain Deluged by Huge Influx of Immigrants," *The Independent* (London, (July 25, 2000)), as reported in *Migration News* 7, 8 (2000) at http://migration.ucdavis.edu/mn/archive_mn/aug_2000-11mn.html, June 19, 2003.

18. Ibid.

19. This price is a reported fee for crossing from the Western Sahara to the Canary Islands. Passage across the Strait to the mainland of Spain has been recorded to cost anything from US$500 to US$1000 per person. See *Migration News* 7, 6 (June 2000) at http://migration.ucdavis.edu/mn/archive_mn/june_2000-13mn.html, 6/19/03 and "Italy, Spain: Illegal Immigrants" *Migration News* 5, 10 (October 1998) at http://migration.ucdavis.edu/mn/archive_mn/oct_1998-13mn.html, June 19, 2003.

20. "Spain, Italy" *Migration News* 9, 5 (May 2002) at http://migration.ucdavis.edu/mn/archive_mn/may_2002-12mn.html, June 19, 2003.

21. David White, "Spain 2000: Tolerance Under Challenge," *Financial Times* (June 13, 2000) as reported in *Migration News* 7, 7 (July 2000) at http://migration.ucdavis.edu/mn/archive_mn/jul_2000-11mn.html, 6/19/03.

22. One of the greatest missing pieces in truthfully reporting the exports of Morocco is the hashish trade. Few if any recognized economic analysts are willing to figure in this enormous cash crop when calculating Morocco's GNP. The result is a gaping hole in the reality of Moroccan economics.

23. The Economist Intelligence Unit's Country Report, Spain, reported that Spain's unemployment rate for the fourth quarter of 2001 was 9.5%. In stark contrast, the Economist Intelligence Unit's Country Report Morocco reported that Morocco's unemployment rate for the fourth quarter of 2001 was 20.3%. See "Country Report: Spain," Economist Intelligence Unit (London, *The Economist*, February 2002) 6 and "Country Report: Morocco," Economist Intelligence Unit (London, *The Economist,* May 2002) 6.

24. "Spain and Portugal: Legalization," *Migration News* 7, 3 (March 2000), at http://migration.ucdavis.edu/mn/archive_mn/mar_2000-11mn.html, June 19, 2003.

25. Ibid. This is true despite the fact that it is estimated by the Association of Moroccan Immigrant Workers in Spain that the wages earned by many Moroccans in Spain are nearly ten times what they could earn in their own country. See *Migration News* 5, 10 (October 1998) at http://migration.ucdavis.edu/mn/archive_mn/oct_1998-13mn.html, June 19, 2003.

26. This was a common complaint made by a number of Spanish interviewees, particularly in the south of Spain and in Barcelona.

27. "Spain, Italy," *Migrant News* 7, 11 (November 2000) at http://migration.ucdavis.edu/mn/archive_mn/nov_2000-13mn.html, June 19, 2003.

28. "Italy, Spain," *Migrant News* 8, 7 (July 2001) at http://migration.ucdavis. edu/mn/archive_mn/jul_2001-11mn.html, June 19, 2003.
29. "Italy, Spain: Illegal Immigrants," as cited in *Migration News* 5, 10 (October 1998) at http://migration.ucdavis.edu/mn/archive_mn/oct_1998-13mn.html, June 19, 2003.
30. It should be noted that among the interviewees cited in this chapter, Maria José Benitez Rojas is one such convert.
31. Interview conducted by Javier Martos Cando, recorded by the author, September 22, 2000 in Quatravitas, and translated from the original Spanish by Deborah Avery.
32. This particular interviewee would only speak anonymously. Interview conducted by Muhammad Oukili and recorded by the author, June 21, 2000, in Fes and translated from the original Derija by Kamal Mzoughi.
33. Interview conducted and recorded by the author, September 27, 2000 in Barcelona, and translated from the original Spanish by Deborah Avery.
34. Interview conducted by Kamal Hassani, recorded by the author, July 7, 2000 in Guelmim and translated from the original Derija and Hassania by Salmam Benzidi.
35. Interview conducted and recorded by the author, September 19, 2000 in Seville, and translated from the original Spanish by Deborah Avery.
36. Interview conducted by Muhammad Oukili and recorded by the author, June 20, 2000, in Fes and translated from the original Derija by Kamal Mzoughi.
37. Interview conducted and recorded by the author, September 30, 2000 in Madrid, and translated from the original Spanish by Peter Huijing.
38. Interview conducted by Kamal Hassani, recorded by the author, July 1, 2000 in Fes, and translated from the original Derija by Salmam Benzidi.
39. Interview conducted and recorded by the author, September 30, 2000, and translated from the original Spanish by Deborah Avery.
40. Interview conducted by Hamid Nouamani, recorded by the author, June 27, 2000 in Ouaoumana, and translated from the original Tamazight by Sadik Rddad.
41. Interview conducted and recorded by the author, September 16, 2000 in Seville, and translated from the original Spanish by Deborah Avery.
42. Interview conducted by Muhammad Oukili, recorded by the author June 13, 2000 in Fes, and translated from the original Derija by Kamal Hassani and the author.
43. A recent report conducted in May 2002 by the Instituto Opina "found that 60% of Spaniards link immigration to increased crime, and two-thirds said that they believed Spaniards were becoming less tolerant of immigrants." See "Spanish Official Says E. U. states should get tough on Traffickers," *Agence France Presse* (May 16, 2002), as cited in *Migration News* 9, 6 (June 2002) at http://migration.ucdavis.edu/mn/archive_mn/jun_2002-11mn. html, June 19, 2003.
44. Ibid. The writer for Agence Press stated, "In 1999 there were 312 arrests for trafficking in illegal laborers, while between January and May of 2002, over 2000 people were arrested for trafficking."
45. This policy resulted in the legalization of approximately 200,000 previously undocumented workers, many of whom were Moroccan. See "Spain,

Italy" as cited in *Migration News* 7, 6 (June 2000) at http://migration. ucdavis.edu/mn/archive_mn/jun_2000-13mn.html, June 19, 2003.

46. Interview conducted and recorded by the author in Las Norias de Daza (El Ejido) on October 18, 2000, and translated from the original Spanish by Deborah Avery.

47. "3-Cultures Foundation Opened by Shimon Peres," *New York Times* (September 9, 1998): A10.

48. I. William Zartman and William Mark Habeeb, eds., *Polity and Society in Contemporary North Africa* (Boulder, CO: Westview Press, 1993), 242.

49. Ian O. Lesser and Ashley J. Tellis, *Strategic Exposure: Proliferation Around the Mediterranean* (Santa Monica: RAND, 1996), 32.

50. "Italy, Spain," as cited in *Migration News* 8, 7 (July 2001) at http:// migration.ucdavis.edu:80/archive_mn/jul_2001-11mn.html, June 19, 2003.

51. "E. U.: Enlargement, Population," as cited in *Migration News* 7, 6 (June 2000) at http://migration.ucdavis.edu:80/mn/archive_mn/jun_2000-09mn.html, June 19, 2003.

Chapter 6   The Future of Sustainable Diplomacy

1. Gwynn Edwards and Ken Haas, *Flamenco!* (London: Thames and Hudson, 2000), 27.

2. Ibid., 52.

3. Will Kirkland, *Gypsy Cante: Deep Song of the Caves* (San Francisco: City Light Books, 1999), 14.

4. Edwards and Haas, *Flamenco!*, 27.

5. David J. Wellman, *Sustainable Communities* (New York: World Council of Churches, 2001), 22.

6. Ibid., 24.

7. Ibid.

8. One of the more recent articles on this subject can be found in the New York Times, entitled, "Amnesty Accuses Spain of Racism in its Treatment of Immigrants," *New York Times* (April 19, 2002): A5.

9. Larry L. Rasmussen, *Earth Community, Earth Ethics* (Maryknoll, NY: Orbis, 1996), 172–173.

10. "Respect" in this context is taken from the Haitian use of the word. In this sense, respect is better spelled "re-spect," meaning to look again with the willingness to see something or someone in a new light. Haitian respect is also intrinsically mutual—everyone must benefit or no one will benefit.

11. Examples of this phenomenon span the globe, from the Balkans, to India, to the Levant.

12. For more on the subject of modern tribalism in a globalized world, see Benjamin R. Barber, *Jihad vs. McWorld: How Globalism and Tribalism are Reshaping the World* (New York: Ballentine Books, 1996).

13. While it is clear that King Mohammed VI intends to modernize his country, including its political system, the majority of power in Morocco still remains with the King. Mohammed's ability to work toward a transition to a truly constitutional monarchy will in many ways determine the ultimate success or failure of his leadership. Regarding Spain's most recent human

rights abuses, see Amnesty International's report, issued in April 2002, or read Emma Daily's article, entitled "Amnesty Accuses Spain of Racism in Its Treatment of Immigrants," *New York Times* (April 19, 2002): A5.

14. Rasmussen, *Earth Community, Earth Ethics*, 286–287.

15. Wellman, *Sustainable Communities*, 24.

16. Brian Swimme and Thomas Berry, *The Universe Story: From the Primordial Flaring Forth to the Ecozoic Era* (San Francisco: Harper-San Francisco, 1992), 243, as cited in Rasmussen, *Earth Community, Earth Ethics*, 30.

17. From a synopsis of Rasmussen's *Earth Community Earth Ethics*, found in Wellman, *Sustainable Communities*, 26.

18. Ibid., 57.

19. Rasmussen, *Earth Community, Earth Ethics*, 129.

20. Douglas Johnston and Cynthia Sampson, eds., *Religion, the Missing Dimension of Statecraft* (New York: Oxford, 1994).

21. See Vandana Shiva, *Monocultures of the Mind: Perspectives on Biodiversity and Biotechnology* (London: Zed Books, 1993).

22. Julio de Santa Ana, *Sustainability and Globalization* (Geneva: World Council of Churches Publications, 1998), 14.

23. Ibid., 16.

24. Ibid., 19.

25. Farid Esack, *Qur'an, Liberation and Pluralism: An Islamic Perspective on Interreligious Solidarity Against Oppression* (London: Oneworld Publications/Oxford, 1998), 158.

26. de Santa Ana, *Sustainability and Globalization*, 19.

27. A Mousem is a traditional feast honoring the life of a salih (some Western translations would equate the term with the Christina notion of "saint"). The celebration is the culmination of a pilgrimage most often to the site of the salih's tomb, where often sacrifices are performed, followed by a feast. This is a time when those who have made the pilgrimage petition the salih or the surviving members of the salih's family for invocations on their behalf in exchange for offerings. A mousem arguably harkens back to pre-Islamic North African traditions, but it is considered by the people who celebrate them to be an integral part of their own Islamic practice and identity. See Dale F. Eikelman, *Moroccan Islam: Tradition and Society in a Pilgrimage Center* (Austin University of Texas Press, 1976), 7, 42, 66, 84–85, 112–113, 171–178.

28. The word *"romeria"* translates from Spanish as "pilgrimage." Like the *mousem*, the *romeria* is a popular (some would say folkloric) expression of Christianity that most likely has its roots in pre-Christian tradition. *Romerias* remain highly popular in many parts of Spain today. A *romeria* involves a group of followers, usually of a virgin, who make a pilgrimage to the site of her image, to petition her for help, and to celebrate her life. Some *romerias* are so popular as to defy normal expectations. The feast of the Rocio, held every spring in Andalucia, transforms the village of the virgin, normally the home of less than 500 people, through welcoming over a million visitors. The Rocio is a spring virgin, associated with fecundity, in sharp contrast to the images of the weeping virgins of Holy Week. Many Roma as well as non-Roma Spaniards travel to the *romeria* of Rocio by oxcart, some from as far away as points in northern Spain. Upon arrival, extravagant feasting and dancing is the order of the day. The line between

Christian worship and pre-Christian worship is blurred. It should be noted that the feast of Rocio is one of largest gatherings in Spain for the entire year. See *El Rocio, Fe y Alegria de un Pueblo* (Granada: Editorial Andalucia de Ediciones Anel, 1981).

Appendix

1. These questions are drawn directly from the work of Daniel T. Spencer, who created and defined the term "Ecological Location" in his book *Gay and Gaia: Ethics, Ecology and the Environment* (Charlotte, OH: The Pilgrim Press, 1996).
2. Subjects in Spain were asked about the land of Spain, and those in Morocco were asked about the land of Morocco.
3. This is a reference to the original peoples of each country. The Spanish subjects were asked about the Iberians of Spain and the Moroccans about the Imazighn of Morocco. These are very general titles to describe far more complex populations.
4. In this case, each group was being asked about the other, i.e. How do you as a Moroccan see yourself as connected to the people living in Spain?
5. The Moroccan interviews were tape-recorded in Fes, Douiet, Ifrane, Ait Oualal, Khenifra, Ouaoumana, Essaouira, Tiznit, Ait Abdellah, Amelen, Tighmi, Tafroute, and Guelmim from June 7–July 24, 2000. The Spanish interviews were tape-recorded in Barcelona, Madrid, Seville, Valdezufre, Cuatrovitas, Aracena, Fuenteheridos, Las Cabezas de San Juan, Arcos de la Frontera, Lebrija, Algeciras, Alanis, and El Ejido from September 8–October 23, 2000. The interviews in Morocco were conducted in Derija, Tamazight, Tachelhite, and English. The interviews in Spain were conducted in Castilian but include interviews with Basque and Catalan speakers.

SELECTED BIBLIOGRAPHY

Abu-Rabi, Ibrahim M. "Concept of the 'Other' in Modern Arab Thought: from Muhammad 'Abdu to Abdallah Laroui." *Islam and Christian–Muslim Relations* 8, no. 1 (1997): 85–97.

Agwan, A. R. "Towards an Ecological Consciousness." *American Journal of Islamic Social Sciences* 10, no. 2 (Summer 1993): 238–248.

Alaoui, Mohammed Ben El Hassan. *La Coopération entre L'Union Européenne et Les Pays du Maghreb*. Paris: Editions Nathan, 1994.

Ali, Ahmed. *Al-Qur'an: A Contemporary Translation*. Princeton, NJ: Princeton University Press, 1993.

Allendesalazar, Jose Manuel. *La Diplomacia Espanola y Marruecos, 1907–1909*. Madrid: Ministerio de Asuntos Exteriores, Agencia Espanola de Cooperacion Internacional, Instituto de Cooperacion con el Mundo Arabe, 1990.

Altemir, Antonio Blanc, ed. *El Mediterráneo: Un Espacio Commún Para La Cooperación, El Desarrollo y El Dialogo Intercultural*. Madrid: Editorial Tecnos, 1999.

An-Na'im, Abdullahi Ahmed. "Islam, Islamic Law and the Dilemma of Cultural Legitimacy for Universal Human Rights." In C. Welch and V. Leary, eds. *Asian Perspectives on Human Rights*. Boulder, CO: Westview Press, 1990.

al-'Arabi, Muhyi-D-Din. *The Voyage of No Return*. Cambridge, UK: The Islamic Texts Society, 2000.

———. *Divine Governance of the Human Kingdom*. Louisville, KY: Fons Vitae, 1997.

———. *The Bezels of Wisdom*. Mahwah, NJ: The Paulist Press, 1980.

———. *Shajarat al-Kawn*. Translated by A. Jeffreys. Lahore, Pakistan: Aziz Publishers, 1980.

———. *The Wisdom of the Prophets*. Aldsworth, Gloucestershire: Beshara Publications, 1975.

———. *Sufis of Andalusia: The Ruh al-Quds and al-Durrat al Fakhirah*. Aldsworth, Gloucestershire: Beshara Publications, 1971.

Arberry, Arthur J. *The Koran Interpreted*. New York: Collier Books, 1955.

Asad, Talal. *Geneologies of Religion: Discipline and Reasons of Power in Christianity and Islam*. Baltimore, MD: Johns Hopkins University Press, 1993.

Aubet, Maria Eugenia. *The Phoenicians and the West: Politics, Colonies and Trade*. New York: Cambridge University Press, 1993.

Baker, Rob, ed. "The Tree of Life Issue." *Parabola* 14, no. 3 (August 1989): 4–83.

Baker, Rob and Gray Henry, eds. *Merton and Sufism: The Untold Story—A Complete Compendium*. Louisville, KY: Fons Vitae, 1999.

Ballesteros, Angel. *Estudio Diplomatico Sobre Ceuta y Melilla*. Cordoba, Argentina: M. Lerner Editora, 1989.

Barber, Benjamin R. *Jihad vs. McWorld: How Globalism and Tribalism are Reshaping the World*. New York: Ballentine Books, 1996.

Basetti-Sani, Guilio O. F. M. *Mohammed et Saint Francois*. Ottawa: Commissariat de Terre-Sainte, 1959.

Bayes, Jane H. and Nayereh Tohidi, eds. *Globalization, Gender, and Religion: The Politics of Women's Rights in Catholic and Muslim Contexts*. New York: Palgrave, 2001.

Ben-Ami, Issachar. *Saint Veneration Among the Jews in Morocco*. Detroit: Wayne State University Press, 1998.

Benseddik, Nacera. "Entre Femme dans le Maghreb Ancien." *AWAL: Cahiers D'Etudes Berbères* 20 (1999): 113–150.

Berry, Wendell. *The Gift of the Good Land*. San Francisco: North Point Press, 1981.

Bielefeldt, Heiner. "Secular Human Rights: Challenge and Opportunity to Christians and Muslims." *Islam and Christian-Muslim Relations* 7, no. 3 (1996): 311–325.

Biggar, Nigel, Jamie S. Scott, and William Schweiker, eds. *Cities of Gods: Faith, Politics, and Pluralism in Judaism, Christianity and Islam*. Westport: Greenwood Press, 1986.

Birch, Bruce C. and Larry L. Rasmussen. *The Predicament of the Prosperous*. Philadelphia: Westminster Press, 1978.

Bonmatí, José Fermín. *Españoles en el Magreb, Siglos XIX y XX*. Madrid: MAPFRE, 1992.

Borelli, John. "The Tree of Life in Hindu and Christian Theology." In R. Masson, ed. *The Pedagogy of God's Image* (Chico, CA: Scholar's Press, 1982): 173–190.

Bormans, Maurice. *Guidelines for Dialogue Between Christians and Muslims*. Mahwah, NJ: Paulist Press, 1990.

Bounds, Elizabeth M. *Coming Together/Coming Apart: Religion, Community, and Modernity*. New York, NY: Routledge, 1997.

Bowles, Paul. *Their Heads are Green and Their Hands are Blue: Scenes From the Non-Christian World*. New York: Echo Press, 1984.

Brett, Michael. *Ibn Khaldun and the Medieval Maghrib*. Brookfield: Ashgate/Variorum, 1999.

Brett, Michael and Elizabeth Fentress. *The Berbers*. Cambridge, MA: Blackwell, 1996.

Bromley, Daniel W., ed. *Making the Commons Work: Theory, Practice and Policy*. San Francisco: ICS Press, 1992.

Brown, Irving. *Deep Song: Adventures with Gypsy Songs and Singers in Andalusia and Other Lands with Original Translations*. New York: Harper and Brothers Publishers, 1929.

Brown, Lester R., ed. *The State of the World 2001*. New York: W. W. Norton, 2001.

Brown, Stuart E. *Meeting in Faith: Twenty Years of Christian-Muslim Conversations Sponsored by the World Council of Churches*. Geneva, Switzerland: WCC Publications, 1989.

Brueggemann, Walter. *Genesis*. Atlanta: John Knox Press, 1982.

Burckhardt, Titus. *La Civilización Hispano-Árabe*. Madrid: Alianza Editorial, 1999.

Callahan, William J. *The Catholic Church in Spain: 1875–1998*. Washington DC: The Catholic University of America Press, 2000.

Calvert, Albert F. *Moorish Remains in Spain*. New York: John Lane Company, 1906.

Cánovas, Ignacio Alcaraz. *Entre España y Marruecos: Testimonio de una Epoca: 1923–1975*. Madrid: Editorial Catriel, 1999.

Castañón, Luz García. *Moros y Cristianos en las Narraciones Infantiles Árabes y Españolas*. Madrid: Ediciones de la Tore, 1995.

Cate, Patrick O'Hair. *Each Other's Scripture: The Muslim's View of the Bible, The Christian's View of the Qur'an*. Ann Arbor, MI: University of Michigan Press, 1974.

Cavalli-Sforza, L. Lucia, Paolo Menozzi, and Alberto Piazza. *The History and Geography Of Human Genes*. Princeton, NJ: Princeton University Press, 1994.

Centre for the Study of Islam and Christian–Muslim Relations. *European Muslims and Christian-Muslim Relations*. Birmingham: Centre for the Study of Christian–Muslim Relations, 1981.

———. *Attitudes to Islam In Europe: An Anthology of Muslim Views*. Birmingham: Centre for the Study of Islam and Christian-Muslim Relations, 1980.

Chew, Sing C. *World Ecological Degradation: Accumulation, Urbanization, and Deforestation 3000 B. C.—A. D. 2000*. New York: Altamira Press, 2001.

Chittick, William C. *The Self-Disclosure of God: Principles in Ibn al-'Arabi's Cosmology*. New York: State University of New York Press, 1998.

———. *Imaginal Words: Ibn al-'Arabi and the Problem of Religious Diversity*. Albany: State University of New York Press, 1994.

Choudhury, Masudul A. "Social Choice in an Islamic Economic Framework." *American Journal of Islamic Social Sciences* 8, no. 2 (1997): 259–274.

Christian, William A., Jr. *Local Religion in Sixteenth-Century Spain*. Princeton: Princeton University Press, 1981

Clover, Frank M. *The Late Roman West and the Vandals*. Brookfield, VT: Variorum, 1993.

Cobb, John B., Jr. *Sustaining the Common Good: A Christian Perspective on the Global Economy*. Cleveland, OH: The Pilgrim Press, 1994.

Cobb, John B., Jr. *Sustainability: Economics, Ecology & Justice*. Maryknoll, NY: Orbis, 1992.

Collins, Roger. *Law, Culture and Regionalism in Early Medieval Spain*. Brookfield, VT: Variorum, 1992.

Combs-Schilling, M. E. *Sacred Performances: Islam, Sexuality and Sacrifice*. New York: Columbia University Press, 1989.

Cornell, Vincent J. *Realm of the Saint: Power and Authority in Moroccan Sufism*. Austin: University of Texas Press, 1998.

Cragg, Kenneth. *The Call of the Minaret*. Maryknoll, NY: Orbis, 1985.

Crapanzano, Vincent. *Tuhami: Portrait of a Moroccan*. Chicago: University of Chicago Press, 1980.

Cudsi, Alexander S. and Ali E. Hillal Dessouki, eds. *Islam and Power*. London: Croom Helm, 1981.

Courbage, Youssef and Philippe Fargues. *Christians and Jews Under Islam*. New York: I. B. Taurus, 1998.

Dabashi, Hamid. *Authority in Islam: From the Rise of Muhammad to the Establishment of the Umayyads*. New Brunswick: Transaction Publishers, 1989.

de la Serna, Alfonso. *Al Sur de Tarifa*. Madrid: Marcel Pons, 2001.

Demorlaine, Madeleine. *Un Seul Dieu, Tous Freres: Jalons pour un Dialogue entre Christiens et Musilmans*. Paris: Chalet, 1986.

Denffer, Ahman von. *Christians in the Qur'an and the Sunna: An Assessment from the Sources to Help Define Our Relationship*. N. P.: The Islamic Foundation, 1979.

Denny, Frederick Mathewson. *An Introduction to Islam*. New York: Macmillan Publishing Company, 1994.

Derrida, Jacques. *The Politics of Friendship*. New York: Verso, 1997.

Derry, Duncan R. et al., *World Atlas of Geology and Mineral Deposits*. New York: John Wiley and Sons, 1980.

Diamond, Jared. *Guns, Germs and Steel: The Fates of Human Societies*. New York: W. W. Norton, 1999.

Djait, Hichem. *Europe and Islam: Cultures and Modernity*. Berkeley: University of California Press, 1985.

Dooling, Dorothea M., ed. "The Tree of Life," includes an array of essays on sacred trees in many different religions and folkloric stories. *Parabola* 14 (Fall 1989): 11–79.

Douglas, William A., Carmelo Urza, Linda White, and Joseba Zulaika, eds. *Basque Cultural Studies, Basque Studies Program Occasional Papers Series no. 5*. Reno: University of Nevada, 1999.

Driessen, Henk. *On the Spanish–Moroccan Frontier: A Study in Ritual, Power and Ethnicity*. New York, NY: Berg Publishers Ltd., 1992.

Dufour-Kowalska, Gabriella. *L'Arbre de Vie et la Croix: Sur L'Imagination Visionnaire*. Geneve: Editions du Tricorno, 1985.

Dumbrell, William J. "Genesis 1–3, Ecology, and the Dominion of Man." *Crux* 21, no. 4 (December 1985): 16–26.

Edwards, Gwynne and Ken Haas. *Flamenco!* London: Thames and Hudson, 2000.

Edwards, John. *Religion and Society in Spain, c. 1492*. Brookfield, VT: Variorum, 1996.

Eickelman, Dale F. *Knowledge and Power in Morocco: The Education of a Twentieth Century Notable*. Princeton: Princeton University Press, 1985.

———. *Moroccan Islam: Tradition and Society in a Pilgrimage Center*. Austin, TX: University of Texas Press, 1976.

Ellingham, Mark, ed. *The Rough Guide to Morocco*. London: Penguin Books, 1997.

Endress, Gerhard. *An Introduction to Islam*. New York: Columbia University Press, 1988.

Entelis, John P., ed. *Islam, Democracy and the State in North Africa*. Indianapolis, IN: Indiana University Press, 1997.

Esack, Farid. *Qur'an, Liberation and Pluralism: An Islamic Perspective on Interreligious Solidarity Against Oppression*. London: One World/Oxford, 1998.

Escámez, Anna Quiñones. *Derecho e Inmigracion: El Repudio Islamico en Europa*. Barcelona: Fundacion "LA CAIXA," 2000.

Esposito, John L. *Islam and Politics*. Syracuse, NY: Syracuse University Press, 1987.

Estevéz, Juan Clemente Rodriguez. *El Alminar de Isbiliya: La Giralda en sus Origenes (1184–1198)*. Sevilla: Colección Giralda, 1998.

Ezzeddin, Ibrahim and Denys Johnson-Davies, trans. *Forty Hadith*. Cambridge, UK: The Islamic Texts Society, 1997.

———. *Forty Hadith Qudsi*. Cambridge, UK: The Islamic Texts Society, 1997.

Fahad, Obaidullah. "Islamic Diplomacy—The Basic Concepts." *Hamard Islamicus* 15 (Spring 1992): 75–93.

Fernea, Elizabeth Warnock. *In Search of Islamic Feminism: One Woman's Global Journey*. New York: Doubleday, 1998.

———. *A Street in Marrakech: A Personal Encounter with the Lives of Moroccan Women*. Garden City, NY: Anchor Books, 1980.

Fernea, Elizabeth Warnock and Robert A. Fernea. *The Arab World: Personal Encounters*. New York: Anchor Books, 1985.

Fiorenza, Elisabeth Schussler. *The Book of Revelation: Justice and Judgement*. Philadelphia: Fortress, 1985.

Frymer-Kinsky, Tykva. *In the Wake of the Goddess*. New York: Free Press, 1992.

Gabus, Jean Paul. *Islam et Christianisme en Dialogue*. Paris: Cerf, 1982.

Garvey, Geoff and Mark Ellingham, eds. *The Rough Guide to Andalucia*. London: Penguin, 1997.

Gatje, Helmut. *The Qur'an and Its Exegesis: Selected Texts with Classical and Modern Muslim Interpretations*. Berleley: University of California Press, 1976.

Geertz, Clifford. *Islam Observed: Religious Development in Morocco and Indonesia*. Chicago: University of Chicago Press, 1968.

Gilhus, Ingvild Saelid. "The Tree of Life and the Tree of Death: A Study of Gnostic Symbols." *Religion* 17 (October 1987): 337–353.

Gillespie, Richard. *Spain and the Mediterranean: Developing a European Policy Towards the South*. New York: St. Martin's Press, 2000.

Gillespie, Richard, ed. *The Euro-Mediterranean Partnership: Political and Economic Perspectives*. Portland, OR: Frank Cass, 1997.

Gillespie, Richard, Fernando Rodrigo, and Jonathan Story, eds. *Democratic Spain: Reshaping External Relations in a Changing World*. London: Routledge, 1995.

Gleave, L. and E. Kermeli, eds. *Islamic Law: Theory and Practice*. London: I. B. Tauris Publishers, 1997.

Gold, Peter. *Europe or Africa?: A Contemporary Study of the Spanish North African Enclaves of Ceuta and Melilla*. Liverpool: Liverpool University Press, 2000.

Goldziher, Ignaz. *Introduction to Islamic Theology and Law*. Princeton: Princeton University Press, 1981.

Gonzales, Valérie. *Beauty and Islam: Asthetics in Islamic Art and Architecture*. New York: I. B. Tauris, 2001.

Gottlieb, Roger S., ed. *This Sacred Earth: Religion, Nature, Environment*. New York: Routledge, 1996.

Green, Jerrold D. and Cynthia Tindell. *Towards NAFTA: A North African Free Trade Agreement?* Santa Monica, CA: Rand, 1995.

Gunton, Colin E. *The One, the Three and the Many: God, Creation and the Culture of Modernity*. Cambridge: Cambridge University Press, 1993.

Haddad, Yvonne Yazbeck. "Islamist Depictions of Christianity in the Twentieth Century: The Pluralism Debate and the Depiction of the Other." *Islam and Christian-Muslim Relations* 7, no. 1 (1996): 75–93.

Haddad, Yvonne Yazbeck and Wadi Z. Haddad, eds. *Christian-Muslim Encounters*. Gainesville, FL: University Press of Florida, 1995.

Hallman, David G., ed. *Ecotheology: Voices from the South and North*. Geneva, Switzerland: WCC Publications, 1994.

Harding, Susan Friend. *Remaking Ibieca: Rural Life in Aragon Under Franco*. Chapel Hill: University of North Carolina Press, 1984.

———. *Tribe and Society in Rural Morocco*. Portland, OR: Frank Cass, 2000.

Hart, David M. *The Ait 'Atta of Southern Morocco: Daily Life and Recent History*. Cambridge, UK: Middle East and North African Studies Press, Ltd., 1984.

Harvey, L. P. *Islamic Spain: 1250 to 1500*. Chicago: University of Chicago Press, 1992.

Hashmi, Sohail Humayun. "Is There an Islamic Ethic of Humanitarian Intervention?" *Ethics and International Affairs* 7 (1993): 55–93.

———. "Toward an Islamic Ethics of International Relations." *American Journal of Islamic Social Sciences* 10 (Spring 1993): 88–95.

Haynes, Jeff. *Religion and Politics in Africa*. London: Zed Books, 1996.

Heather, Peter. *The Visigoths From the Migration Period to the Seventh Century: An Ethnographic Perspective*. San Marino: The Boydell Press, 1999.

Heckt, Paul. *The Wind Cried: An American's Discovery of the World of Flamenco*. Westport, CT: The Bold Strummer Ltd., 1994.

Hessel, Dieter T., ed. *Theology for Earth Community: A Field Guide*. Maryknoll, NY: Orbis, 1996.

Hiebert, Theodore. *The Yahwist's Landscape: Nature and Religion in Early Israel*. New York, NY: Oxford University Press, 1996.

Hillel, Daniel J. *Out of the Earth: Civilization and the Life of the Soil*. Berkeley, CA: University of California Press, 1994.

Hillel, Daniel J. *Rivers of Eden: The Struggle for Water and the Quest for Peace in the Middle East*. New York, NY: Oxford University Press, 1994.

Hirtenstein, Stephen. *The Unlimited Mercifier: The Spiritual Life and Thought of Ibn 'Arabi*. Ashland, OR: White Cloud Press, 1999.

Hoffmann, Stanley. *Duties Beyond Borders: On the Limits and Possibilities of Ethical International Politics*. Syracuse, NY: Syracuse University Press, 1981.

Hooper, John. *The New Spaniards*. London: Penguin Books, 1995.

Hualde, Jose Ignacio, Joseba A. Lakarra, and R. L. Trask. *Towards a History of The Basque Language: Current Issues in Linguistic Theory, 131*, Philadelphia: John Benjamins Publishing Company, 1996.

Hughes, J. Donald. *Ecology in Ancient Civilizations*. Albuquerque NM: University of New Mexico Press, 1975.

Hunwick, John D. *Religion and National Integration in Africa: Islam, Christianity, and Politics in the Sudan and Nigeria*. Northwestern University Press, 1992.

Idel, Moshe and Bernard McGinn, eds. *Mystical Union in Judaism, Christianity and Islam: An Ecumenical Dialogue*. New York: Continuum, 1996.

Johnston, Douglas and Cynthia Sampson, eds. *Religion, The Missing Dimension of Statecraft*. New York: Oxford University Press, 1994.

Joseph, Suad and Barbara L. K. Pillsbury. *Muslim-Christian Conflicts: Economic, Political, and Social Origins*. Boulder, CO: Westview Press, 1978.

Junco, José Alvarez and Adrian Shubert, eds. *Spanish History since 1808*. New York: Oxford University Press, 2000.

Kassis, Hannah E. "Christian Misconceptions of Islam." In M. Briemberg, ed. *It Was, It Was Not*. Vancouver BC: New Star Books, 1992.

———, comp. *A Concordance of the Qur'an*. Berkeley: University of California Press, 1983.

Kathir Ad-Damishqi, Imam Imâduddin Abdul-Fida Isma'îl Ibn. *Stories of the Prophets*. Riyadh: Darussalam, 1999.

Keddie, Nikki R. "Ideology, Society and the State in Post-Colonialist Muslim Societies." In A. Kimmens, ed. *Islamic Politics*. New York: H. W. Wilson, 1991.

Kelsay, John. "Divine Command Ethics in Early Islam: Al-Shafi'i and the Problems of Guidance." *The Journal of Religious Ethics* 22, no. 1 (Spring 1994): 101–126.

———. "Islamic Law and Ethics (Bibliography)." *The Journal of Religious Ethics* 22 (Spring 1994): 95–99.

———, ed. *Just War and Jihad: Historical and Theoretical Perspectives on War and Peace in Western and Islamic Traditions*. New York, NY: Greenwood Press, 1991.

Kennedy, Hugh. *Muslim Spain and Portugal: A Political History of al-Andalous*. New York: Longman, 1996.

Keohane, Alan. *The Berbers of Morocco*. London: Hamish Hamilton, 1991.

Khan, Mohammed A. Muqtedar. "Islam as an Ethical Tradition of International Relations." *Islam and Muslim–Christian Relations* 8, no. 2 (Spring 1994): 177–192.

Kimball, Charles. *Striving Together: A Way Forward in Christian–Muslim Relations*. Maryknoll, NY: Orbis, 1991.

Kirkland, Will. *Gypsy Cante: Deep Songs of the Caves*. San Francisco: City Lights Books, 1999.

Knitter, Paul F. *One Earth Many Religions: Multifaith Dialogue & Global Responsibility*. Maryknoll, NY: Orbis, 1996.

Lalaguna, Juan. *A Traveller's History of Spain*. New York: Interlink Books, 1996.

Lancel, Serge. *Carthage: A History*. Cambridge, MA: Blackwell, 1995.

———. *Marruecos: Islam y Nacionalismo*. Madrid: MAPFRE, 1994.

Laroui, Abdallah. *The History of the Maghrib: An Interpretive Essay*. Princeton: Princeton University Press, 1977.

Larrazabal, Ramon Salas. *El Protectorado de Espana en Marruecos*. Madrid: Editorial MAPFRE, 1992.

Le Beau, Bryan F. and Menachem Mor, eds. *Religion in the Age of Exploration: The Case of Spain and New Spain*. Omaha: Creighton University Press, 1996.

Lerner, Steve. *Eco-Pioneers: Practical Visionaries Solving Today's Environmental Problems*. Cambridge, MA: The MIT Press, 1997.

Lesser, Ian O. *Southern Europe and the Maghreb: U. S. Interests and Policy Perspectives*. Santa Monica, CA: Rand, 1996.

Lewis, Bernard. *Islam and the West*. Oxford: Oxford University Press, 1993.

———. *The Muslim Discovery of Europe*. New York: W. W. Norton, 1982.

———, ed. *Inmigración Magrebí en España: El Retorno de los Moriscos*. Madrid: MAPFRE, 1993.

López, Bernabé, ed. *España-Magreb, Siglo XXI*. Madrid: MAPFRE, 1992.

McDaniel, Jay B. *With Roots and Wings: Christianity in an Age of Ecology And Dialogue*. Maryknoll, NY: Orbis, 1995.

McFague, Sallie. *The Body of God: An Ecological Theology*. Minneapolis: Fortress Press, 1993.

McLane, Merrill. *East from Granada: Hidden Andalusia and Its People*. Cabin John, MD: Carderock Press, 1996.

McNeill, J. R. *The Mountains of the Mediterranean World: An Environmental History*. New York: Cambridge University Press, 1992.

Magne, Jean. *From Christianity to Gnosis and from Gnosis to Christianity: An Itinerary Through the Texts to and from the Tree of Paradise*. Atlanta: Scholars Press, 1993.

Maguire, Daniel C. and Larry Rasmussen. *Ethics for a Small Planet: New Horizons on Population, Consumption, and Ecology*. Albany, NY: State University of New York, 1998.

Maimonides, Moses. *The Guide for the Perplexed*. New York: Dover Publications, 1956.

Mairota, Paola, John B. Thornes, and Nichola Geeson, eds. *Atlas of Mediterranean Environments in Europe: The Desertification Context*. New York: John Wiley and Sons, 1998.

Mandel, Robert. *The Changing Face of National Security: A Conceptual Analysis*. Westport, CT: Greenwood Press, 1994.

———. *Conflict Over the World's Resources: Background, Trends, Case Studies, and Considerations for the Future*. Westport, CT: Greenwood Press, 1988.

———. *Irrationality in International Confrontation*. Westport, CT: Greenwood Press, 1987.

Mander, Jerry and Edward Goldsmith, eds. *The Case Against the Global Economy*. San Francisco: Sierra Club Books, 1996.

Mann, Vivian, ed. *Morocco: Jews and Art in a Muslim Land*. New York: The Jewish Museum, 2000.

Mann, Vivian, Thomas Glick, and Jerrilynn Dodds, eds. *Convivencia: Jews, Christians and Muslims in Medieval Spain*. New York: The Jewish Museum, 1992.

Marias, Julian. *Understanding Spain*. Ann Arbor, MI: The University Of Michigan Press, 1992.

Markoe, Glenn. *Phoenicians*. London: British Museum Press, 2000.

Marquez, Teodoro Falcon. *Torres de Almenara del Reino de Granada en Tiempos de Carlos III*. Sevilla: Junta de Andalucia, 1989.

Martin, Susan F., ed. *World Migration Report 2000*. United Nations: United Nations Publications, 2000.

Masson, Denise. *L'Eau, le Feu, la Lumi: La Bible, Le Coran et les Traditions Monothistes*. Paris: Descle de Brouwer, 1985.

Matar, Nabil. *Turks, Moors and Englishmen in the Age of Discovery*. New York: Columbia University Press, 1999.

Maxwell-Hyslop, K. R. "The Assyrian Tree of Life: A Western Branch." In J. V. S. Megaw, ed. London: Thames and Hudson, *To Illustrate the Monuments*, 1976, 263–276.

Memeley, Anne. *Tournaments of Value: Sociability and Hierarchy in a Yemeni Town*. Toronto: University of Toronto Press, 1996.

Mernissi, Fatima. *Islam and Democracy: Fear of the Modern World*. New York: Addison-Wesley Publishing Company, 1992.

———. *The Veil and the Male Elite: A Feminist Interpretation of Women's Rights in Islam*. New York: Addison-Wesley Publishing Company, 1987.

Meyerson, Mark and Edward D. English, eds. *Christians, Muslims and Jews in Medieval and Early Modern Spain*. Notre Dame, IN: University of Notre Dame Press, 1999.

Mies, Maria and Vandana Shiva. *Ecofeminism*. London: Zed Books, 1993.

Mitri, Tarek, ed. *Religion, Law and Society: A Christian–Muslim Discussion*. Geneva, Switzerland: WCC Publications, 1995.

Moha, Edouard. *Las Relaciones Hispano-Marroquies*. Malaga, Spain: Editorial Algazara, 1992.

Mojuetan, Benson Akutse. *History and Underdevelopment in Morocco: The Structural Roots of Conjecture*. Munster: Lit Verlag, 1995.

Moltmann, Jurgen. *The Source of Life: The Holy Spirit and the Theology of Life*. Minneapolis, MN: Fortress Press, 1997.

———. *God in Creation: A New Theology of Creation and the Spirit of God*. New York: Harper and Row, 1985.

———. *The Future of Creation*. Philadelphia: Fortress Press, 1979.

Morgenthau, Hans J. *Politics Among Nations: The Struggle for Power and Peace*. New York: Alfred A. Knopf, 1978.

Mortimer, Edward. *Faith and Power: The Politics of Islam*. New York: Random House, 1982.

Moscati, Sabatino, ed. *The Phoenicians*. New York: Abbeville Press, 1988.

Muñoz, Gema Martín. *Islam, Modernism and the West*. New York: I. B. Tauris, 1999.

Munson, Henry, Jr. *Religion and Power in Morocco*. New Haven, CT: Yale University Press, 1993.

Murata, Sachiko and William C. Chittick. *The Vision of Islam*. St. Paul, MN: Paragon House, 1994.

Nasr, Seyyed Hossein. *Religion and the Order of Nature*. New York: Oxford University Press, 1996.

———. *An Introduction to Islamic Cosmological Doctrines*. Albany: State University of New York Press, 1993.

Nazarea, Virginia D. *Ethnoecology: Situated Knowledge/Located Lives*. Tucson: University of Arizona Press, 1999.

Neuman, Stephanie G., ed. *International Relations Theory and the Third World*. New York: St. Martin's Press, 1998.

Nicol, George G. "The Threat and the Promise (Symbolism of the Tree of Life)." *Expository Times* 94 (February 1993): 136–139.

Nielsen, J. S. "Muslims in Europe: History Revisited or a Way Forward?" *Islam and Christian–Muslim Relations* 8, no. 2 (1997): 135–143.

Nogales, José Luis Sánchez, ed. *De la Frontera al Encuentro: Actas del Tercer Curso ≪Cristanos y Musulmanes≫ (Granada, 30 de Junio a 4 de Julio de 1997)*. Granada: Centro de Investigación sobre las Relaciones Interreligiosas (CIRI) Facultad de Teología de Granda, 1997.

Nomani, Farhad and Ali Rahnema. *Islamic Economic Systems*. London: Zed Books, 1994.

Nooteboom, Cees. *Roads to Santiago: Detours and Riddles in the Lands and History of Spain*. New York: Harcourt and Brace, 1992.

O'Dowd, Liam and Thomas Wilson, eds. *Borders, Nations and States: Frontiers of Sovereignty in the New Europe*. Brookfield, VT: Avebury, 1996.

O'Fahey, R. S. *Enigmatic Saint: Ahmad Ibn Idris and the Idrisi Tradition*. Evanston, IL: Northwestern University Press, 1990.

Ogilvie, Sheilagh C. and Marcus Cerman, eds. *European Proto-Industrialization*. Cambridge, UK: Cambridge University Press, 1996.

Olupona, J. *Religion and Peace in a Multi-Faith Nigeria*. Ile-Ife, Nigeria: Obafemi Awolono University, 1992.

Pagels, Elaine. *Adam, Eve and the Serpent*. New York, NY: Vintage Books, 1989.

Parker, Richard B. *North Africa: Regional Tensions and Strategic Concerns*. New York: Praeger, 1987.

Pascon, Paul. *Capitalism and Agriculture in the Haouz of Marrakesh*. New York: KPI Ltd., 1986.

Pearson, Birger A. "Notes and Observations: 'She Became a Tree'—A Note to CG II, 4: 89, 25–26." *The Harvard Theological Review* 69 (1976): 413–415.

Pennell, C. R. *Morocco since 1830: A History*. New York: New York University Press, 2000.

Peters, F. E. *Judaism, Christianity and Islam: The Classical Texts and Their Interpretation, vol. I: From Covenant to Community*. Princeton, NJ: Princeton University Press, 1990.

Peters, F. E. *Judaism, Christianity and Islam: The Classical Texts and Their Interpretation, vol. II: The Word and the Law and the People of God*. Princeton, NJ: Princeton University Press, 1990.

Phillips, Jack, "The Cross as the Tree of Life in the Coptic Church." *Coptic Church Review* 10 (Summer 1989): 53–54.

Phipps, William E. *Muhammad and Jesus: A Comparison of the Prophets and Their Teachings*. New York: Continuum, 1996.

Pino, Domingo del. *La Ultima Guerra con Marruecos: Ceuta y Melilla*. Barcelona, Spain: Argos Vergara, 1983.

Pohren, D. E. *The Art of Flamenco*. Morón de la Frontera (Sevilla): Finca Espartero, 1972.

Primavesi, Ann. *From Apocalypse to Genesis: Ecology, Feminism and Christianity*. Minneapolis: Fortress Press, 1991.

Rahbar, Daud. *God of Justice: A Study in the Ethical Doctrine of the Qur'an*. Leiden: E. J. Brill, 1960.

———. *Major Themes of the Qur'an*. Minneapolis, MN: Bibliotheca Islamica, 1989.

Rahman, Fazlur. *Islam*. Chicago: University of Chicago Press, 1979.

Rasmussen, Larry L. *Earth Community Earth Ethics*. Maryknoll, NY: Orbis, 1996.

Rasmussen, Larry L. *Moral Fragments & Moral Community: A Proposal for Church in Society*. Minneapolis, MN: Fortress Press, 1993.

Reilly, Bernard F. *The Contest of Christian and Muslim Spain: 1031–1157*. Cambridge, MA: Blackwell, 1995.

——. *The Medieval Spains*. New York: Cambridge University Press, 1993.

Reinhart, A. Kevin. *Before Revelation: The Boundaries of Muslim Moral Thought*. Albany, NY: State University of New York Press, 1995.

Renard, John, S. J., ed. *Windows on the House of Islam: Muslim Sources on Spirituality and Religious Life*. Berkeley, CA: University of California Press, 1998.

——. *Seven Doors to Islam: Spirituality and the Religious Life of Muslims*. Berkeley, CA: University of California Press, 1996.

——, ed. *Ibn 'Abbad of Ronda: Letters on the Sufi Path*. New York, NY: Paulist Press, 1986.

Richard, Pablo. *Apocalypse: A People's Commentary on the Book of Revelation*. Maryknoll, NY: Orbis, 1995.

Rippin, Andrew and Jan Knappert, eds. *Textual Sources for the Study of Islam*. Chicago: University of Chicago Press, 1986.

Robinson, Neal. *Christ in Islam and Christianity*. New York: State University of New York Press, 1991.

Rogers, Eamonn, ed. *Encyclopedia of Contemporary Spanish Culture*. New York: Routledge, 2002.

Roloff, Jurgen. *The Revelation of John: A Continental Commentary*. Minneapolis, MN: Fortress Press, 1995.

Roth, Norman. *Jews, Visigoths and Muslims in Medieval Spain: Cooperation and Conflict*. New York: E. J. Brill, 1994.

Rubin, Barnett R. *Cases and Strategies for Preventative Action*. New York: The Century Foundation Press, 1998.

Rubio, Carolina Hernández, ed. *Estructura Economica del Mundo: II. El Magreb y Oriente Medio*. Madrid: San Lorenzo de El Escoral, 1999.

Ruedy, John, ed. *Islamism and Secularism in North Africa*. New York: St. Martin's Press, 1996.

Ruether, Rosemary Radford. *Gaia and God: An Ecofeminist Theology of Earth Healing*. San Francisco: Harper Collins, 1992.

Ruiz, Teofilo. *Crisis and Continuity: Land and Town in Late Medieval Castile*. Philadelphia, PA: University of Pennsylvania Press, 1994.

Rumi, Melana Celaleddin. *Divan-I Kebir*. Walla Walla, WA: Current, 1995.

Sachedina, Abdulaziz. "Islamic Theology of Christian-Muslim Relations." *Islam and Christian–Muslim Relations* 8, no. 1 (1997): 27–38.

el Sadawi, Nawal. *The Nawal el Sadawi Reader*. London: Zed Books, 1997.

Salisbury, Joyce E. *Iberian Popular Religion: 600 B. C. to 700 A. D: Celts, Romans and Visigoths*. New York: The Edwin Mellon Press, 1985.

Santamaria, Ramiro. *Ifni y Sahara: La Guerra Ignorada*. Madrid: Ediciones Dyrsa, 1984.

Scharper, Stephen Bede. *Redeeming the Time: A Political Theology of the Environment*. New York: Continuum, 1997.

Schimmel, Annemarie. *Deciphering the Signs of God: A Phenomenological Approach to Islam*. Albany: State University of New York Press, 1994.

Schreiner, Claus, ed. *Flamenco: Gypsy Dance and Music from Andalusia*. Portland, OR: Amadeus Press, 1990.

Seale, Moris S. *Qur'an and Bible: Studies in Interpretation and Dialogue*. London: Croom Helm, 1978.

Sells, Michael A., ed. *Early Islamic Mysticism: Sufi, Qur'an, Mi'Raj, Poetic and Theological Writings*. New York, NY: Paulist Press, 1996.

Serageldin, Ismail. "The Justly Balanced Society: One Muslim's View." In D. Beckman, ed. *Friday Morning Reflections at the World Bank*. Washington DC: Seven Locks Press, 1991.

Shaida, S. A. "Islamic Ethics: Some Theoretical Questions." *Journal of Objective Studies* 1 (1947).

Shiva, Vandana. *Biopiracy: The Plunder of Nature and Knowledge*. Boston, MA: South End Press, 1997.

——. *Monocultures of the Mind: Perspectives on Biodiversity and Biotechnology*. London: Zed Books, 1993.

——. *The Violence of the Green Revolution: Third World Agriculture, Ecology and Politics*. London: Zed Books, 1991.

——. *Staying Alive: Women, Ecology and Development*. London: Zed Books, 1989.

Shriver, Donald W., Jr. *An Ethic for Enemies: Forgiveness in Politics*. New York: Oxford University Press, 1995.

Shubert, Adrian. *A Social History of Modern Spain*. New York: Routledge, 1996.

Siddiqui, Ataullah. *Christian–Muslim Dialogue in the Twentieth Century*. New York: St. Martin's Press, 1997.

Silva, Antonio Torremocha. *Algeciras entre la Cristianidad y el Islam*. Algeciras (Cadiz): Impresur, S. L., 1994.

Simpson, James. *Spanish Agriculture: The Long Siesta, 1765–1965*. New York: Cambridge University Press, 1995.

Sivan, Emmanuel. *Radical Islam: Medieval Theology and Modern Politics*. New Haven: Yale University Press, 1985.

Slomp, Jan. *A Short Bibliography on Muslim–Christian Relations*. N. P.: Consultative Committee on Islam in Europe of the Conference of European Churches, 1982.

Smith, Jackie, Charles Chatfield, and Ron Pagnucco, eds. *Transnational Social Movements and Global Politics: Solidarity Beyond the State*. Syracuse: Syracuse University Press, 1997.

Spencer, Daniel T. *Gay and Gaia: Ethics, Ecology, and the Erotic*. Cleveland, OH: The Pilgrim Press, 1996.

Stavridis, Stelios, Theodore Couloumbis, Thanos Veremis, and Neville Waites, ed. *The Foreign Policies of the European Union's Mediterranean States and Applicant Countries in the 1990s*. New York: St. Martin's Press, 1999.

Stewart, Charles F. *The Economy of Morocco: 1912–1962*. Cambridge, MA: Harvard University Press, 1967.

Strange, John. "The Idea of Afterlife in Ancient Israel: Some Remarks on the Iconography in Solomon's Temple." *Palestine Exploration Quarterly* (January–June 1985): 35–40.

Stocking, Rachel L. *Bishops, Councils, and Consensus in the Visigothic Kingdom, 589–633*. Ann Arbor: University of Michigan Press, 2000.

Swearingen, Will D. *Moroccan Mirages: Agrarian Dreams and Deception 1912–1986*. Princeton, NJ: Princeton University Press, 1987.

Swearingen, Will D. and Abdellatif Bencherifa, eds. *The North African Environment at Risk*. Boulder, CO: Westview Press, 1996.

Swimme, Brian and Thomas Berry. *The Universe Story: From the Primordial Flaring Forth to the Ecozoic Era, a Celebration of the Unfolding of the Cosmos*. San Francisco, CA: Harper Collins, 1992.

Taha, 'Abdulwahid Dhanun. *The Muslim Conquest and Settlement of North Africa and Spain*. New York: Routledge, 1989.

Thompson, Kenneth W. *Ethics, Functionalism, and Power in International Politics*. Baton Rouge, LA: Louisiana State University, 1979.

Thompson, Leonard L. *The Book of Revelation: Apocalypse and Empire*. New York: Oxford University Press, 1990.

Timm, Roger E. "The Ecological Fallout of Islamic Creation Theology." In Mary Evelyn Tucker, ed. *Worldviews and Ecology*. Lewisberg, PA: Bucknell University Press, 1993, 83–95.

Tortella, Gabriel. *The Development of Modern Spain: An Economic History of the Nineteenth and Twentieth Centuries*. Cambridge, MA: Harvard University Press, 2000.

townes, emilie M. ed. *A Troubling in My Soul: Womanist Perspectives on Evil and Suffering*. Maryknoll, NY: Orbis, 1996.

———. *In a Blaze of Glory: Womanist Spirituality as Social Witness*. Nashville, TN: Abingdon Press, 1995.

Truxillo, Charles A. *By the Sword and the Cross: The Historical Evolution of the Catholic World Monarchy in Spain and the New World, 1492–1825*. Westport, CT: Greenwood Press, 2001.

Tsevat, Matitiahu, "The Two Trees in the Garden of Eden." In B. Mazar, ed. *Nelson Glueck Memorial Volume*, 40–43 (English summary of Hebrew text, 19).

Twomey, Lesley. *Faith and Fanaticism: Religious Fervor in Early Modern Spain*. Brookfield: Ashgate, 1997.

Us-Zaman, Waheed. "Doctrinal Position of Islam Concerning Inter-State and International Relations." *Hamdard Islamicus* 9 (Spring 1986): 81–91.

Van Oudenaren, John. *Employment, Economic Development and Migration in Southern Europe and the Maghreb*. Santa Monica, CA: Rand, 1996.

Van Sertima, Ivan, ed. *Golden Age of the Moor*. New Brunswick, NJ: Transaction Publishers, 1993.

Vassberg, David E. *Land and Society in Golden Age Castile*. New York: Cambridge University Press, 1984.

Vertovek, Steven and Ceri Peach, eds. *Islam in Europe: The Politics of Religion and Community*. New York: St. Martin's Press, 1997.

Vilar, Juan Bta. and Ramón Lourido. *Relaciones entre Espana y el Maghreb: Siglos XVII y XVIII*. Madrid: MAPFRE, 1994.

Von Rad, Gerhard. *Genesis*. Philadelphia: Westminster Press, 1972.

Waardenburg, Jaques. "Critical Issues in Muslim–Christian Relations: Theoretical, Practical, Dialogical, Scholarly." *Islam and Christian–Muslim Relations* 8, no. 1 (March 1997): 9–26.

Wackernagel, Mathis and William Rees. *Our Ecological Footprint: Reducing Human Impact on Earth*. Gabriola Island, BC, Canada: New Society Publishers, 1996.

Walker, H. J. and W. E. Grabau. *The Evolution of Geomorphology: A Nation-By-Nation Summary of Development*. New York: John Wiley and Sons, 1993.

Walker, Williston, Richard A. Norris, David W. Lotz, and Robert Handy. *A History of the Christian Church, Fourth Edition*. New York: Charles Scribner's Sons, 1985.

Wallace, Mark I. *Fragments of the Spirit: Nature, Violence, and the Renewal of Creation*. New York: Continuum, 1996.

Waltz, Susan E. *Human Rights and Reform: Changing the Face of North African Politics*. Berkeley, CA: University of California Press, 1995.

Washabaugh, William. *Flamenco: Passion, Politics and Popular Culture*. Washington, DC: Berg, 1996.

Watson, Andrew M. *Innovaciones en la Agricultura en los Prieros Tiempos del Mundo Islámico*. Granada: Universidad de Granada, 1998.

Watt, William Montgomery. *A Short History of Islam*. Oxford: Oneworld, 1996.

———. *Islamic Political Thought*. Edinburgh: Edinburgh University Press, 1968.

Wellman, David J. *Sustainable Communities*. New York: World Council of Churches, 2001.

Westermark, Edward Alexander. *Ritual and Belief in Morocco, vols. I and II*. New Hyde Park, NY: University Books, 1968.

Wiarda, Howard J. and Margaret MacLeish Mott. *Catholic Roots and Democratic Flowers: Political Systems in Spain and Portugal*. Westport, CT: Praeger, 2001.

Wingate, Andrew. *Encounter in the Spirit: Muslim–Christian Meetings in Birmingham*. Geneva: World Council of Churches, 1988.

Wolf, Eric R., ed. *Religious Regimes and State Formation: Perspectives from European Ethnology*. Albany, NY: State University of New York Press, 1991.

Wolfram, Herwig. *The Roman Empire and Its Germanic Peoples*. Berkeley: University of California Press, 1997.

Wosk, Yosef J. *Two Trees Planted in the Midst of an Enigmatic Garden: A Four-Dimensional Study of Genesis 2: 8–9*. Thesis (Ph.D.): Boston University, 1992.

Yahya, Dahiru. *Morocco in the Sixteenth Century: Problems and Patterns in African Foreign Policy*. Harlow, UK: Longman, 1981.

Zartman, I. William and William Mark Habeeb. *Polity and Society in Contemporary North Africa*. Boulder, CO: Westview Press, 1993.

Zebiri, Kate. *Muslims and Christians Face to Face*. London: Oneworld/Oxford Publications, 1997.

Zehili, Wahba Moustapha. "Dispositions Internationales Relatives la Guerre, Justifies au Regard de L'Islam, et Leurs Aspects Humains Caractristiques." In P. Viaud, ed. *Les Religions et la Guerre: Judaisme, Christianisme, Islam*. Paris: Cerf, 1991.

INDEX